Department of Transport
Scottish Development Department
Welsh Office

# METHOD OF MEASUREMENT FOR ROAD AND BRIDGE WORKS

Davies, Middleton & Davies Ltd.
Cartwright House
Lower Bristol Road
Twerton
BATH BA2 1EU

Her Majesty's Stationery Office
London

© Crown copyright 1977
First published 1971
Second Edition 1977
Reprinted 1982

ISBN 0 11 550439 7

# CONTENTS

| Part I | Definitions | Page 1 |
|---|---|---|

| Part II | General Principles | 3 |
|---|---|---|

| Part III | Rules for Preparing Bills of Quantities | 5 |
|---|---|---|
| | Preambles to Bill of Quantities | 8 |

| Part IV | Units and Methods of Measurement | 11 |
|---|---|---|

| | Section 1 | Preliminaries | 12 |
|---|---|---|---|
| | | Temporary Accommodation | 12 |
| | | Operatives for the Engineer | 14 |
| | | Traffic Safety and Control (Traffic Safety Measures) | 14 |
| | | Temporary Diversion of Traffic | 15 |
| | | Vehicles for the Engineer | 16 |
| | | Progress Photographs | 17 |

| | Section 2 | Site Clearance | 18 |
|---|---|---|---|

| | Section 3 | Hedges | 20 |
|---|---|---|---|
| | | Hedges | 20 |
| | | Excavation in Rock and Reinforced Concrete | 20 |

| | Section 4 | Fencing | 22 |
|---|---|---|---|
| | | Fences, Gates and Stiles | 22 |
| | | Safety Fences | 24 |
| | | Excavation in Rock and Reinforced Concrete | 26 |

| | Section 5 | Drainage and Service Ducts | 28 |
|---|---|---|---|
| | | Sewers, Drains, Piped Culverts and Ducts (excluding French Drains) | 28 |
| | | French Drains | 31 |
| | | Manholes, Catchpits, Drawpits and Gullies | 33 |
| | | Intercepting Ditches | 35 |
| | | Headwalls and Outfall Works | 36 |
| | | Excavation in Rock and Reinforced Concrete | 36 |
| | | Reinstatement of Pavement | 37 |

|  |  |  |
|---|---|---|
| | Soft Spots and Other Voids | 38 |
| | Supports left in Excavation | 39 |
| | Drainage and Ducts in Bridges, Viaducts and Other Structures | 39 |
| | Filling to Pipe Bays on Bridges | 40 |
| **Section 6** | **Earthworks** | **42** |
| | Excavation | 43 |
| | Deposition of Fill | 46 |
| | Disposal of Material | 47 |
| | Imported Fill | 48 |
| | Compaction of Fill | 49 |
| | Soft Spots and Other Voids | 51 |
| | Supports left in Excavation | 51 |
| | Soiling | 52 |
| | Grassing | 52 |
| | Completion of Formation | 53 |
| | Lining of Watercourses | 54 |
| | Clearing of Existing Ditches | 54 |
| **Section 7** | **Roadworks Overall Requirements** | **56** |
| **Section 8** | **Sub-base and Roadbase** | **57** |
| **Section 9** | **Flexible Surfacing** | **59** |
| **Section 10** | **Concrete Pavement** | **62** |
| **Section 11** | **Kerbs and Footways** | **64** |
| | Kerbing, Channelling and Edging | 64 |
| | Excavation in Rock and Reinforced Concrete | 65 |
| | Footways | 65 |
| **Section 12** | **Traffic Signs and Road Markings** | **67** |
| | Traffic Signs | 67 |
| | Road Markings | 68 |
| | Road Studs | 69 |
| | Marker Posts | 70 |
| | Excavation in Rock and Reinforced Concrete | 70 |
| **Section 13** | **Piling for Structures** | **72** |
| | Piling Plant | 72 |
| | Precast Concrete Piles | 73 |
| | Cast-in-Place Piles | 75 |
| | Reinforcement for Cast-in-Place Piles | 77 |
| | Steel Sheet Piles | 78 |
| | Steel Bearing Piles | 81 |

| | | |
|---|---|---|
| Section 14 | **Formwork for Structures**<br>Formwork<br>Patterned Profile Formwork | 84<br>84<br>85 |
| Section 15 | **Steel Reinforcement for Structures** | 88 |
| Section 16 | **Concrete for Structures**<br>In situ Concrete<br>Precast Members<br>Treatment to Concrete Faces after the<br>Striking of Formwork | 90<br>90<br>91<br><br>92 |
| Section 17 | **In situ Post-tensioned Prestressing for Structures** | 94 |
| Section 18 | **Steelwork for Structures**<br>Fabrication of Steelwork<br>Erection of Steelwork<br>Corrugated Steel Structures | 96<br>96<br>97<br>98 |
| Section 19 | **Protection of Steelwork against Corrosion** | 100 |
| Section 20 | **Waterproofing for Structures** | 102 |
| Section 21 | **Bridge Bearings** | 104 |
| Section 22 | **Metal Parapets** | 105 |
| Section 23 | **Movement Joints for Structures**<br>Movement Joints to Bridge and Viaduct Decks<br>Movement Joints other than to<br>Bridge and Viaduct Decks | 106<br>106<br><br>107 |
| Section 24 | **Brickwork for Structures** | 109 |
| Section 25 | **Masonry for Structures** | 111 |
| Section 26 | (Not taken up) | 113 |
| Section 27 | **Testing**<br>Pile Testing<br>Practical Tests of Concrete for Structures<br>Testing Precast Concrete Members for Structures<br>Grouting Trials<br>Welding and Flame Cutting Procedure Trials<br>Test of Bearings | 114<br>114<br>115<br>115<br>116<br>117<br>117 |
| Section 28 | (Not taken up) | 119 |
| Section 29 | **Accommodation Works and Works for<br>Statutory Undertakers** | 120 |

# PART I

# Definitions

## Definitions

**1** In this document entitled Method of Measurement for Road and Bridge Works (hereinafter referred to as 'the Method of Measurement') unless the context otherwise requires, the following words and expressions shall have the meanings hereby respectively assigned to them, that is to say:

(a) 'Conditions of Contract' means the Conditions of Contract referred to in the Tender;

(b) words and expressions to which meanings are assigned in the Conditions of Contract have the same meanings in the Method of Measurement;

(c) words and expressions to which meanings are assigned in the Specification and Drawings referred to in the Conditions of Contract have the same meanings in the Method of Measurement;

(d) 'Bill of Quantities' means a list of items giving brief identifying descriptions and estimated quantities of work comprised in the execution of the works to be performed;

(e) 'Daywork' means the method of valuing work on the basis of the time spent by the workmen, the materials used and the plant employed;

(f) Items designated as 'Provisional' and required to be priced by the Tenderer are items for which the quantities of work to be executed cannot be determined with the same degree of accuracy as other items but for which it is deemed necessary to make provision;

(g) 'Preliminary Item' means an item in respect of works and or general obligations and risks antecedent to or involved in the execution of the Contract and which is set out either in a 'Preliminaries' section of the Bill of Quantities and relating to the Works as a whole or under a 'Special Preliminaries' sub-heading relating to a particular section of the Works;

(h) 'Other Structure' means a structure that is neither a bridge nor a viaduct;

(i) 'Finishings' means the miscellaneous surfacings, furniture and ancillary features that are added to any part of a structure;

(j) 'Existing Ground Level' means the level of the ground before any work under the Contract has been carried out.

# PART II

## General Principles

## General Principles 4

**1** The Bill of Quantities is intended in the first instance to give information upon which tenders can be obtained. When a Contract has been entered into, the function of the priced Bill of Quantities is to provide for the valuation of the work executed.

**2** The sub-headings and item descriptions used in the Bill of Quantities identify the work covered by the respective items, but the exact nature and extent of the work to be performed is to be ascertained by reference to the Drawings, Specification and Conditions of Contract as the case may be, read in conjunction with the matters listed against the relevant marginal headings 'Item coverage' in Part IV hereof. The rates and prices to be inserted in the Bill of Quantities are to be considered as the full inclusive rates and prices for the finished work covered by the respective items and as covering all labour, materials, temporary work, plant, overhead charges and profit, as well as the general liabilities, obligations and risks arising out of the Conditions of Contract.

**3** Each item description is to be consistent with and be compounded from one or more of the descriptive features listed in the numbered sequence of Groups in the various Sections of Part IV hereof, as many of these Groups or Features being used as may be necessary to identify the work required, but not more than one feature from any one Group may be represented in any one item description. In the event of any change or addition to the Specification for Road and Bridge Works dated 1976 it will be necessary to ascertain whether the Method of Measurement adequately covers for such changes or additions. Where it does not so cover, the Method of Measurement shall be amended as appropriate. (See footnote to Part III Preambles to Bill of Quantities.)

**4** Unless expressly stated otherwise in the Contract the Bill of Quantities is to contain all those items compounded in accordance with the foregoing paragraph 3 required to comprise the Works (apart from Provisional Sums and Prime Cost Items which may be required).

**5** Where items are included in a Bill of Quantities for work to be executed or goods, materials or services to be supplied by a Nominated Sub-Contractor then separate items are to be provided for:
(i) labours in connection therewith in the form of a lump sum;
(ii) all other charges and profit in connection therewith in the form of a percentage.

# PART III

# Rules for Preparing Bills of Quantities

# Rules for Preparing Bills of Quantities

**1** The Bill of Quantities is to be sub-divided into separate sections as appropriate in the following sequence:
Preliminaries
Roadworks
Each Bridge, Viaduct or Other Structure
Service Areas
Maintenance Compounds
Accommodation Works
Works for Statutory Undertakers
Testing
Dayworks

**2** The items in the Bill of Quantities for Roadworks, Service Areas, Maintenance Compounds, Accommodation Works and Works for Statutory Undertakers are to be grouped as appropriate under the following section-headings:
Site Clearance
Hedges
Fencing
Drainage and Service Ducts
Earthworks
Sub-base and Roadbase
Flexible Surfacing
Concrete Pavement
Kerbs and Footways
Traffic Signs and Road Markings

**3** The items in the Bill of Quantities for each bridge, viaduct or other structure are to be grouped as appropriate under the following construction headings:

| *Bridges and Viaducts* | *Other Structures* |
|---|---|
| Special Preliminaries | Special Preliminaries |
| Special Foundations | Special Foundations |
| Substructure-End Supports | Structure |
| Substructure-Intermediate Supports | Finishings |
| Superstructure | Testing |
| Finishings | |
| Testing | |

Where more than one distinct type of construction is involved for either substructure or superstructure as, for example, might occur in a river crossing with approach spans and river spans of different structural forms, or in the suspended spans of some cantilever and suspended span designs, the substructure and superstructure construction headings of the Bill of Quantities may be further subdivided under appropriate headings.

**4** Work to be executed within and below:
(a) the area covered by non-tidal open water (such as rivers, canals, lakes and the like) at the highest water level shown on the Drawings; or

## Rules for Preparing Bills of Quantities

(b) the area covered by tidal water at the high tide level shown on the Drawings;

shall be identified in the Bill of Quantities under a sub-heading or in the item description as falling within (a) or (b) above and measured separately.

**5** The matters set out under the heading 'Preambles to Bill of Quantities' (1–8) hereafter are always to be included. Additional numbered Preambles may be included as necessary.

# Preambles to Bill of Quantities

**General Directions**

1 In this Bill of Quantities the sub-headings and item descriptions identify the work covered by the respective items, but the exact nature and extent of the work to be performed is to be ascertained by reference to the Drawings, Specification and Conditions of Contract as the case may be read in conjunction with the matters listed against the relevant marginal headings 'Item coverage' in Part IV of the Method of Measurement for Road and Bridge Works published by Her Majesty's Stationery Office in 1977.

The rates and prices entered in the Bill of Quantities shall be deemed to be the full inclusive value of the work covered by the several items including the following, unless expressly stated otherwise:

(i) Labour and costs in connection therewith.
(ii) The supply of materials, goods, storage and costs in connection therewith including waste and delivery to Site.
(iii) Plant and costs in connection therewith.
(iv) Fixing, erecting and installing or placing of materials and goods in position.
(v) Temporary Works.
(vi) The effect on the phasing of the Works of alterations or additions to existing services and mains to the extent that such work is set forth or reasonably implied in the documents on which the tender is based.
(vii) General obligations, liabilities and risks involved in the execution of the Works set forth or reasonably implied in the documents on which the tender is based.
(viii) Establishment charges, overheads and profit.

**Measurement**

2 The measurement of work shall be computed net from the Drawings unless stated otherwise in the Method of Measurement.

The Contractor shall allow in the rates and prices for waste.

**Unpriced Items**

3 Items against which no price or rate is entered shall be deemed to be covered by the other rates and prices in the Bill of Quantities.

**Use of Alternative Authorised Materials or Designs**

4 Where in the Contract a choice of alternative materials or designs is indicated for a given purpose, the description billed and the rates and prices inserted shall be deemed to cover any of the permitted alternative materials or designs which the Contractor may elect to use and all measurement of such work shall be based upon the thinnest alternative construction permitted by the Contract.

## Preambles to Bill of Quantities

**Privately and Publicly Owned Services**

**5** The information in the Contract as to the whereabouts of existing services and mains is believed to be correct but the Contractor shall not be relieved of his obligations under Clause 11 of the Conditions of Contract. The Contractor shall include in his rates and prices for taking measures for the support and full protection of pipes, cables and other apparatus during the progress of the Works and for keeping the Engineer informed of all arrangements he makes with the owners of privately owned services, Statutory Undertakers and Public Authorities as appropriate and for ensuring that no existing mains and services are interrupted without the written consent of the appropriate authority.

**Labours**

**6** Labours in connection with nominated Sub-Contractors shall include:
(a) in the case of work or services executed, for affording the use of existing working space, access, temporary roads, erected scaffolding, working shelters, staging, ladders, hoists, storage, latrines, messing, welfare and other facilities existing on Site and the provision of protection, water, electricity for lighting and clearing away rubbish and debris arising from the work;
(b) in the case of goods, materials or services supplied, for taking delivery, unloading, storing, protecting and returning crates, cartons and packing materials.

**Roadworks Overall Requirements**

**7** The Contractor shall allow in the rates and prices for roadworks for complying with the requirements of the Series No. 700 Roadworks Overall Requirements clauses of the Specification.

**Work Within and Below Non-Tidal Open or Tidal Water**

**8** The Contractor shall allow in the rates and prices for taking measures required to execute the work separately measured as being within and below non-tidal open or tidal water.

For the measurement of work affected by non-tidal open or tidal water the level shown on the Drawings shall be used irrespective of the actual level or area of water encountered in the Works.

**\*Amendment to the Method of Measurement**

For the purpose of this Contract the Method of Measurement referred to in Preamble 1 General Directions is amended in accordance with the pages immediately following.

**\*Footnote**

Where amendments to Part IV of the Method of Measurement are required in accordance with paragraph 3 of the General Principles, this preamble should be the last numbered preamble and inserted immediately prior to the amendments.

# PART IV

## Units and Methods of Measurement

# Section 1: Preliminaries

**Temporary Accommodation**

**Units**

1 The units of measurement shall be:
  (i) erection, servicing, dismantling of temporary accommodation.................item.

**Measurement**

2 The measurement of:
(a) servicing offices for the Engineer until completion of the Works shall be the continuous period from occupation and use of those offices until completion of the whole of the Works certified by the Engineer in accordance with Clause 48(1) of the Conditions of Contract;
(b) servicing offices for the Engineer during the period up to the issue of the Maintenance Certificate shall be the continuous period subsequent to (a) defined above until the issue of the Maintenance Certificate in accordance with Clause 61(1) of the Conditions of Contract.

**Itemisation**

3 Separate items shall be provided for temporary accommodation in accordance with Part II paragraphs 3 and 4 and the following:

| Group | Feature |
|---|---|
| I | 1 Erection.<br>2 Servicing.<br>3 Dismantling. |
| II | 1 Principal offices for the Engineer.<br>2 Principal laboratories for the Engineer.<br>3 Portable offices for the Engineer.<br>4 Portable laboratories for the Engineer.<br>5 Offices and messes for the Contractor.<br>6 Stores and workshops for the Contractor. |
| III | 1 Until completion of the Works.<br>2 During the period up to the issue of the Maintenance Certificate. |

Note: The measurement of the Group III features shall be confined to the servicing of those offices for use by the Engineer specially listed in the Contract as being required to remain for occupation and use until the issue of the Maintenance Certificate.

## Preliminaries

**Erection of Temporary Accommodation**

**Item coverage**

**4** *The items for erection of temporary accommodation shall in accordance with the Preambles to Bill of Quantities General Directions include for:*
 (i) *in the case of accommodation for the Contractor,*
 (a) *everything required by the Contractor.*
 (ii) *in the case of accommodation for the Engineer,*
 (a) *provision of a site for the accommodation;*
 (b) *preparation of site;*
 (c) *foundations, bases and hardstandings;*
 (d) *water, sanitation, heating, power and lighting services;*
 (e) *fences, notice and direction boards;*
 (f) *vehicle access, hardstandings, parking areas and foot paths;*
 (g) *office and laboratory equipment, furnishings, fittings, supplies and initial consumable stores;*
 (h) *telephones, extensions and switchboard separately connected to the Post Office telephone system.*
 *Note: The Employer shall repay to the Contractor the net cost of telephone calls charged to the telephone number or numbers allocated to the Engineer.*

**Servicing Temporary Accommodation**

**Item coverage**

**5** *The items for servicing temporary accommodation shall in accordance with the Preambles to Bill of Quantities General Directions include for:*
 (i) *in the case of accommodation for the Contractor,*
 (a) *everything required by the Contractor.*
 (ii) *in the case of accommodation for the Engineer,*
 (a) *rental charges including telephone rental;*
 (b) *charges for heating, power, lighting and water including water rates and sewage disposal;*
 (c) *depreciation and maintenance of buildings, services, fences, notice and direction boards, vehicle access, parking areas, hardstandings and footpaths;*
 (d) *depreciation and maintenance of office and laboratory equipment, furnishings, fittings and supplies;*
 (e) *cleaning accommodation;*
 (f) *moving and re-establishing portable accommodation as required;*
 (g) *replenishment of consumable stores.*

**Dismantling Temporary Accommodation**

**Item coverage**

**6** *The items for dismantling temporary accommodation shall in accordance with the Preambles to Bill of Quantities General Directions include for:*
 (a) *receiving back from the Engineer and removing equipment, furniture, fittings and supplies off Site;*
 (b) *disconnecting and removing services and sealing off disused services;*
 (c) *demolishing and removing off Site temporary accommodation, vehicle access, hardstandings, parking areas, footpaths, fences, notice and direction boards;*

**Preliminaries** 14

(d) *disposal of surplus material;*
(e) *reinstatement of the sites occupied by temporary accommodation;*
(f) *in the case of accommodation for the Engineer, the credit value of surplus equipment or material which becomes the property of the Contractor and transport and delivery to the Employer of equipment or material which becomes the property of the Employer.*

### Operatives for the Engineer

**Units**

**7** The units of measurement shall be:
   (i) operatives for the Engineer.................day.

**Measurement**

**8** The measurement of operatives for the Engineer shall be a continuous period of four hours or more within any one day during which the operatives services are supplied in accordance with the written order of the Engineer.

**Itemisation**

**9** Separate items shall be provided for operatives for the Engineer in accordance with Part II paragraphs 3 and 4 and the following:

| Group | Feature |
|---|---|
| I | 1 Each type of operative for the Engineer. |

**Operatives for the Engineer**

**Item coverage**

**10** The items for operatives for the Engineer shall in accordance with the Preambles to Bill of Quantities General Directions include for:
(a) *the wages and other emoluments paid to the operative including payment for overtime;*
(b) *the operative working outside the Contractors normal working hours if so required by the Engineer;*
(c) *costs and expenses incurred consequent upon the employment or hiring of the operative;*
(d) *periods of less than four hours which are not measured.*

### Traffic Safety and Control (Traffic Safety Measures)

**Units**

**11** The units of measurement shall be:
   (i) traffic safety and control..................item.

**Itemisation**

**12** Separate items shall be provided for traffic safety and control in accordance with Part II paragraphs 3 and 4 and the following:

| Group | Feature |
|---|---|
| I | 1 Traffic safety and control. |

# Preliminaries

**Traffic Safety and Control**

**Item coverage**

13 *The items for traffic safety and control shall in accordance with the Preambles to the Bill of Quantities General Directions include for:*
   (a) *complying with Chapter 8 of the Traffic Signs Manual published by Her Majesty's Stationery Office and any amendment thereto or where the circumstances of any particular case are not covered submitting proposals for dealing with such situations to the Engineer for approval;*
   (b) *consulting with statutory and other authorities concerned and submitting to the Engineer for his approval a programme showing the proposed scheme of traffic management and furnishing such further details and information as necessitated by the Works or as the Engineer may require;*
   (c) *cleaning, repositioning, covering and removing temporary traffic signs, lamps, barriers and traffic control signals.*

## Temporary Diversion of Traffic

**Units**

14 The units of measurement shall be:
   (i) taking measures for or construction, maintenance, removal of temporary diversion of traffic..............item.

**Itemisation**

15 Separate items shall be provided for temporary diversion of traffic in accordance with Part II paragraphs 3 and 4 and the following:

| Group | Feature |
|---|---|
| I | 1 Taking measures for or construction of temporary diversion of traffic.<br>2 Maintenance of temporary diversion of traffic.<br>3 Removal of temporary diversion of traffic. |
| II | 1 At locations listed in the Schedule.<br>2 At those locations listed in the Schedule but not measured individually. |

**Taking Measures for or Construction of Temporary Diversion of Traffic**

**Item coverage**

16 *The items for taking measures for or construction of temporary diversion of traffic shall in accordance with the Preambles to Bill of Quantities General Directions include for:*
   (a) *obtaining licences;*
   (b) *making arrangements with owners and occupiers of land temporarily required and costs arising therefrom;*
   (c) *preparing, amending and submitting to the Engineer and other interested bodies, Drawings showing proposals and programme including liaison;*
   (d) *preparation of site;*

## Preliminaries

(e) *carriageways, ramps, structures, footways, kerbs, parapets, fencing, road markings and drainage;*
(f) *street lighting to a standard acceptable to the Highway Authority;*
(g) *unless otherwise stated in the Contract, temporary diversions of services, sewers and drains;*
(h) *modifications.*

**Maintenance of Temporary Diversion of Traffic**

**Item coverage**

17 *The items for maintenance of temporary diversion of traffic shall in accordance with the Preambles to Bill of Quantities General Directions include for:*
(a) *maintenance to provide adequately for the traffic flows.*

**Removal of Temporary Diversion of Traffic**

**Item coverage**

18 *The items for removal of temporary diversion of traffic shall in accordance with the Preambles to Bill of Quantities General Directions include for:*
(a) *breaking up diversion;*
(b) *unless otherwise stated in the Contract reinstatement of the site to its previous condition;*
(c) *disposal of surplus material.*

### Vehicles for the Engineer

**Units**

19 The units of measurement shall be:
  (i) vehicles for the Engineer.................vehicle week.

**Measurement**

20 The measurement of vehicles for the Engineer shall be each week or part thereof during which a vehicle is provided.

**Itemisation**

21 Separate items shall be provided for vehicles for the Engineer in accordance with Part II paragraphs 3 and 4 and the following:

| Group | Feature |
|---|---|
| I | 1 Each type of vehicle for the Engineer. |

**Vehicles for the Engineer**

**Item coverage**

22 *The items for vehicles for the Engineer shall in accordance with the Preambles to Bill of Quantities General Directions include for:*
(a) *equipment;*
(b) *taxing for use on public highways and for the carriage of goods and samples;*
(c) *comprehensive insurance covering the Engineer and any driver authorised by him and the carriage of goods and samples;*
(d) *provision of a suitable replacement including equipment when a regular vehicle is unavailable or unserviceable for more than 24 hours;*
(e) *depreciation;*

Preliminaries 17

(f) *maintenance in a roadworthy condition and in conformity with the vehicle manufacturer's recommendations;*
(g) *fuel and oil;*
(h) *keeping clean inside and out*
(i) *collection from Site when the vehicle is returned.*

**Progress Photographs**

**Units**  23 The units of measurement shall be:
    (i) set of progress photographs, set of aerial progress photographs...................number.
    (ii) additional progress photographs, additional aerial progress photographs...................number.

**Measurement**  24 A set of photographs shall comprise such numbers of negatives and prints as described in the Contract taken on any one flight or visit to Site.

Where in any one flight or visit to Site the Engineer orders less than one complete set of photographs, then one set shall be measured.

Where in any flight or visit the Engineer orders progress or aerial photographs in excess of the number in the set then the additional photographs shall be measured and be deemed to include the negative and the same number of prints per negative as those in the set.

**Itemisation**  25 Separate items shall be provided for progress photographs in accordance with Part II paragraphs 3 and 4 and the following:

| Group | Feature |
|---|---|
| I | 1 Set of progress photographs.<br>2 Set of aerial progress photographs.<br>3 Additional progress photographs.<br>4 Additional aerial progress photographs. |

**Progress Photographs, Aerial Progress Photographs, Additional Progress Photographs, and Additional Aerial Progress Photographs**

26 The items for progress photographs, aerial progress photographs, additional progress photographs and additional aerial progress photographs shall in accordance with the Preambles to Bill of Quantities General Directions include for:

**Item coverage**  (a) *delivery of negatives and prints to the Engineer;*
(b) *identification marking on the prints.*

# Section 2: Site Clearance

**Units**

1 The units of measurement shall be:

(i) general site clearance.................hectare.
(ii) demolition of individual or groups of buildings or structures.................item.
(iii) removal of disused sewers or drains................ linear metre.

Note: The taking up of existing carriageways, pavings, kerbs, other than those required to be set aside for re-laying or removing to store, demolition of underground structures, chambers or foundations (including in the case of demolition of buildings or structures, work below and including ground slab and oversite concrete) shall be measured in the measurement of Earthworks.

**Measurement**

2 The measurement of general site clearance shall be the plan area. No deduction shall be made for buildings, structures or carriageways.

**Itemisation**

3 Separate items shall be provided for site clearance in accordance with Part II paragraphs 3 and 4 and the following:

| Group | Feature |
|---|---|
| I | 1 General site clearance. |
| | 2 General site clearance of sections separately indicated. |
| | 3 Demolition of individual or groups of buildings or structures. |
| | 4 Removal of disused sewers or drains of different diameters with 1 metre or less of cover to formation level. |

**General Site Clearance**

4 *The items for general site clearance shall in accordance with the Preambles to Bill of Quantities General Directions include for:*

**Item coverage**

(a) *demolition, breaking up and removal of all walls and superficial obstructions affected by the Works down to existing ground level except as otherwise stated in the Contract;*
(b) *felling of trees;*
(c) *grubbing up and blasting stumps and roots;*
(d) *filling stump and root holes with suitable material from any source and compaction;*

# Site Clearance

(e) uprooting of bushes, small trees and hedges;
(f) credit value of material which becomes the property of the Contractor;
(g) disposal of surplus material;
(h) disconnecting and sealing services;
(i) making good to severed ends of existing fences and walls.

**Demolition of Individual or Groups of Buildings or Structures**

**Item coverage**

5 The items for demolition of individual or groups of buildings or structures shall in accordance with the Preambles to Bill of Quantities General Directions include for:
(a) demolition of the buildings or structures, including blasting if necessary down to existing ground level but excluding the removal of any ground slab or oversite concrete;
(b) credit value of material which becomes the property of the Contractor;
(c) disposal of surplus material;
(d) disconnecting and sealing services.

**Removal of Disused Sewers or Drains**

**Item coverage**

6 The items for removal of disused sewers or drains shall in accordance with the Preambles to Bill of Quantities General Directions include for:
(a) excavation in any material including rock and reinforced concrete and loading into transport, upholding the sides and keeping the earthworks free of water;
(b) preliminary site trials of blasting;
(c) loosening or breaking up by blasting or other means sewer or drain of whatever type of construction together with any bed and haunch or surround;
(d) cutting through reinforcement;
(e) trimming the bottom and sides of excavation and clearing away loose rock;
(f) overbreak and making good;
(g) backfilling with suitable material from any source and compaction;
(h) disposal of surplus material;
(i) sealing the ends of sewers or drains;
(j) backfilling manholes, catchpits and gullies with suitable material from any source and compaction.

# Section 3: Hedges

|  |  |
|---|---|
| **Units** | **Hedges** |
|  | **1** The units of measurement shall be: |
|  | (i) hedges.................linear metre. |
| **Measurement** | **2** The measurement shall be the developed length along the centre line of the hedge. |
| **Itemisation** | **3** Separate items shall be provided for hedges in accordance with Part II paragraphs 3 and 4 and the following: |

| Group | Feature |
|---|---|
| I | 1 Each species of hedge. |
| II | 1 Different spacings of plants. |
| III | 1 Hedges with protective fencing. |

|  |  |
|---|---|
| **Hedges** | **4** *The items for hedges shall in accordance with the Preambles to Bill of Quantities General Directions include for:* |
| **Item coverage** | (a) *excavation and loading into transport, upholding the sides and keeping the earthworks free of water;* |
|  | (b) *excavation from stockpile, loading into transport, and haulage and deposition of top soil in trench;* |
|  | (c) *planting;* |
|  | (d) *protection of plants from injurious weather before and during planting;* |
|  | (e) *fertilising, watering and weeding;* |
|  | (f) *maintenance and replacement;* |
|  | (g) *disposal of surplus material.* |
|  | **Excavation in Rock and Reinforced Concrete** |
| **Units** | **5** The units of measurement shall be: |
|  | (i) excavation in rock, excavation in reinforced concrete .................cubic metre. Measured extra over Hedges. |
| **Measurement** | **6** The measurement shall be the volume of the voids formed by the removal of the rock and reinforced concrete. |
|  | For this measurement, the dimensions of trenches shall be taken as the dimensions shown on the Drawings or ordered by the Engineer. |
| **Itemisation** | **7** Separate items shall be provided for excavation in rock and |

reinforced concrete in accordance with Part II paragraphs 3 and 4 and the following:

| Group | Feature |
|---|---|
| I | 1 Excavation in rock.<br>2 Excavation in reinforced concrete. |

**Excavation in Rock and Reinforced Concrete**

**Item coverage**

8 *The items for excavation in rock and reinforced concrete shall in accordance with the Preambles to Bill of Quantities General Directions include for:*
 (a) *preliminary site trials of blasting;*
 (b) *loosening or breaking up by blasting or other means of unexcavated material before or in the process of excavation;*
 (c) *cutting through reinforcement;*
 (d) *trimming the bottom and sides of excavation and clearing away loose rock;*
 (e) *over break and making good.*

# Section 4: Fencing

**Units**

**Fences, Gates and Stiles**

1 The units of measurement shall be:
  (i) fences, removal and re-erection of existing fences ................. linear metre.
  (ii) gates, stiles, removal and re-erection of existing gates and stiles ................. number.
  (iii) concrete footing to intermediate posts ............ ............ number.

**Measurement**

2 The measurement shall be the developed length along the centre line of the fencing. Temporary fencing required by the Contractor in the discharge of his obligations under Clauses 19 and 22 of the Conditions of Contract shall not be measured. Concrete footing to intermediate posts shall only be measured for those locations shown on the Drawings or ordered by the Engineer.

**Itemisation**

3 Separate items shall be provided for fencing in accordance with Part II paragraphs 3 and 4 and the following:

| Group | Feature |
|---|---|
| I | 1 Each type of fence.<br>2 Concrete footing to intermediate posts of each type of fence.<br>3 Each type of gate.<br>4 Each type of stile.<br>5 Removal and re-erection of each type of existing fence.<br>6 Removal and re-erection of each type of existing gate and stile. |
| II | 1 Fences of different heights.<br>2 Gates of different heights and widths. |
| III | 1 Painted fences or gates. |

**Fences**

**Item coverage**

4 The items for fences shall in accordance with the Preambles to Bill of Quantities General Directions include for:
(a) excavation and loading into transport, upholding the sides and keeping the earthworks free of water;
(b) trimming ground on the line of the fence;
(c) formwork (as Section 14 paragraph 4) to concrete footings;

Fencing

(d) reinforcement (as Section 15 paragraph 5) to concrete footings;
(e) in situ concrete (as Section 16 paragraph 4) in footings to struts and straining posts;
(f) backfilling and compaction;
(g) disposal of surplus material;
(h) preservation of timber;
(i) adjustment of fence to a flowing alignment including additional length posts;
(j) joining to existing fences, gates, hedges and walls;
(k) inspection and maintenance of fencing and gates;
(l) erection and removal of temporary fences and gates;
(m) maintenance of access for owners, tenants and occupiers of adjoining land and patrolling gaps or openings.

**Concrete Footing to Intermediate Posts**

**Item coverage**

5 The items for concrete footing to intermediate posts shall in accordance with the Preambles to Bill of Quantities General Directions include for:
(a) excavation and loading into transport, upholding the sides and keeping the earthworks free of water;
(b) formwork (as Section 14 paragraph 4);
(c) reinforcement (as Section 15 paragraph 5);
(d) in situ concrete (as Section 16 paragraph 4);
(e) backfilling and compaction;
(f) disposal of surplus material.

**Gates and Stiles**

**Item coverage**

6 The items for gates and stiles shall in accordance with the Preambles to Bill of Quantities General Directions include for:
(a) excavation and loading into transport, upholding the sides and keeping the earthworks free of water;
(b) trimming ground at entrance;
(c) formwork (as Section 14 paragraph 4);
(d) reinforcement (as Section 15 paragraph 5);
(e) in situ concrete (as Section 16 paragraph 4);
(f) backfilling and compaction;
(g) disposal of surplus material;
(h) preservation of timber;
(i) posts, hinges, fastenings, stops and locks;
(j) joining to existing fences, hedges and walls;
(k) in the case of new gates and stiles in existing fences, hedges or walls, forming openings and making good.

**Removal and Re-erection of Existing Fence**

**Item coverage**

7 The items for removal and re-erection of existing fence shall in accordance with the Preambles to Bill of Quantities General Directions include for:
(a) breaking up foundations, taking down, cleaning, stacking, protecting and labelling;
(b) storage;
(c) making good to ends of fences, hedges and walls;
(d) replacing any item damaged during the foregoing operations;

## Fencing

(e) fences (*as this Section paragraph 4*).

**Removal and Re-erection of Existing Gate and Stile**
**Item coverage**

**8** *The items for removal and re-erection of existing gate and stile shall in accordance with the Preambles to Bill of Quantities General Directions include for:*
(a) *breaking up foundations, taking down, cleaning, stacking, protecting and labelling;*
(b) *storage;*
(c) *making good to ends of fences, hedges and walls;*
(d) *replacing any item damaged during the foregoing operations;*
(e) *gates and stiles* (*as this Section paragraph 6*).

### Safety Fences

**Units**

**9** The units of measurement shall be:
(i) beams ................... linear metre.
(ii) posts, mounting brackets, terminals, full height anchorages, expansion joint anchorages, connections to bridge parapets, transition pieces, concrete footings to posts ................... number.
(iii) removal of existing safety fences ................... linear metre.

**Definitions**

**10** The term 'beam' shall mean a longitudinal member spanning posts and mounting brackets within the limits defined in paragraph 11 below and shall be deemed to include the terms 'barrier', 'rail', 'safety fence' and 'guardrail' for vehicles.

The term 'mounting bracket' shall be deemed to include the term 'bridge pier or concrete parapet mounting connection'.

**Measurement**

**11** The measurement of beams shall be the developed length along the centre line of the beams or in the case of double sided fences, measured once only along the centre line of the posts between the following points:
(a) the end of each beam type at a connection to bridge parapet or within a transition piece assembly;
(b) the connection of beams to terminals, full height anchorages and expansion joint anchorages.

**12** The measurement of terminals, full height anchorages, expansion joint anchorages and connections to bridge parapets shall be the complete installation including posts and concrete. Mounting brackets and all other posts required between those points defined in paragraph 11 shall be measured and concrete footings to those posts shall only be measured where no other alternative method of fixing is permitted.

**13** The measurement of transition pieces shall be the complete installation.

# Fencing

**Itemisation**

**14** Separate items shall be provided for safety fences in accordance with Part II paragraphs 3 and 4 and the following:

| Group | Feature |
|---|---|
| I | 1 Each type of beam. |
|   | 2 Each type of post. |
|   | 3 Each type of mounting bracket. |
|   | 4 Each type of terminal. |
|   | 5 Each type of full height anchorage. |
|   | 6 Each type of expansion joint anchorage. |
|   | 7 Each type of connection to bridge parapet. |
|   | 8 Each type of transition piece. |
|   | 9 Each type of concrete footing to post. |
|   | 10 Removal of existing safety fence. |

**Beams**

**Item coverage**

**15** *The items for beams shall in accordance with the Preambles to Bill of Quantities General Directions include for:*
*(a) fabrication and preparation;*
*(b) galvanising;*
*(c) attachments, fixings and stiffeners and adjustment of beams to flowing alignment;*
*(d) tensioning and flaring;*
*(e) painting chevrons including background.*

**Posts**

**Item coverage**

**16** *The items for posts shall in accordance with the Preambles to Bill of Quantities General Directions include for:*
*(a) fabrication and preparation;*
*(b) galvanising;*
*(c) preservation of timber;*
*(d) driving or excavation and loading into transport, upholding the sides and keeping the earthworks free of water;*
*(e) backfilling and compaction;*
*(f) disposal of surplus material;*
*(g) post in concrete footing in lieu of driven post where adopted as a fixing by the Contractor including formwork (as Section 14 paragraph 4) reinforcement (as Section 15 paragraph 5) in situ concrete (as Section 16 paragraph 4) and plastic bag;*
*(h) fixing to structures;*
*(i) fixing to beam including spacers.*

**Mounting Bracket**

**Item coverage**

**17** *The items for mounting bracket shall in accordance with the Preambles to Bill of Quantities General Directions include for:*
*(a) fabrication and preparation;*
*(b) galvanising;*
*(c) fixing to structures including adaptor platforms;*
*(d) fixing to beams.*

## Fencing

**Terminal, Full Height Anchorage, Expansion Joint Anchorage, Connection to Bridge Parapet and Transition Piece**

**Item coverage**

18 The items for terminal, full height anchorage, expansion joint anchorage, connection to bridge parapet and transition piece shall in accordance with the Preambles to Bill of Quantities General Directions include for:

(a) fabrication and preparation;
(b) galvanising;
(c) preservation of timber;
(d) driving or excavation and loading into transport, upholding the sides and keeping the earthworks free of water;
(e) formwork (as Section 14 paragraph 4);
(f) reinforcement (as Section 15 paragraph 5);
(g) in situ concrete (as Section 16 paragraph 4);
(h) disposal of surplus material;
(i) fixing to or setting in concrete;
(j) attachments, fixings and stiffeners and adjustment to flowing alignment;
(k) tensioning and flaring;
(l) fixing to bridge parapet;
(m) rounded end section and fishtails;
(n) fixing to beam.

**Concrete Footing to Posts**

**Item coverage**

19 The items for concrete footing to posts shall in accordance with the Preambles to Bill of Quantities General Directions include for:

(a) excavation and loading into transport, upholding the sides and keeping the earthworks free of water;
(b) formwork (as Section 14 paragraph 4);
(c) reinforcement (as Section 15 paragraph 5);
(d) in situ concrete (as Section 16 paragraph 4);
(e) plastic bag;
(f) disposal of surplus material.

**Removal of Existing Safety Fences**

**Item coverage**

20 The items for removal of existing safety fences shall in accordance with the Preambles to Bill of Quantities General Directions include for:

(a) dismantling of beams, breaking up footings, extracting posts, cleaning, protecting, labelling and transporting beams and posts to store off Site nominated by the Engineer;
(b) backfilling post holes with suitable material from any source and compaction;
(c) disposal of surplus material.

### Excavation in Rock and Reinforced Concrete

**Units**

21 The units of measurement shall be:
    (i) excavation in rock, excavation in reinforced concrete................cubic metre. Measured extra over Fencing.

## Fencing

**Measurement**

**22** The measurement shall be the volume of the voids formed by the removal of the rock and reinforced concrete.
For this measurement, the dimensions of post holes shall be taken as the dimensions shown on the Drawings or ordered by the Engineer.

**Itemisation**

**23** Separate items shall be provided for excavation in rock and reinforced concrete in accordance with Part II paragraphs 3 and 4 and the following:

| Group | Feature |
|---|---|
| I | 1 Excavation in rock. |
|   | 2 Excavation in reinforced concrete. |

**Excavation in Rock and Reinforced Concrete**

**Item coverage**

**24** *The items for excavation in rock and reinforced concrete shall in accordance with the Preambles to Bill of Quantities General Directions include for:*
(a) *preliminary site trials of blasting;*
(b) *loosening or breaking up by blasting or other means of unexcavated material before or in the process of excavation;*
(c) *cutting through reinforcement;*
(d) *breakdown of material to comply with the requirements of fill;*
(e) *trimming the bottom and sides of excavation and clearing away loose rock;*
(f) *over break and making good.*

# Section 5: Drainage and Service Ducts

**Definitions**

**1** The Earthworks Outline is defined as the finished earthworks levels and dimensions required by the Contract for the construction of the carriageway, hard shoulder, hard strip, paved area, sub-base material under verge, central reserve and side slope, footway, topsoiling and turfing.

**2** Where the filling between a french drain and an adjacent carriageway, hard shoulder or hard strip is of filter material (contiguous filter material) the Earthworks Outline shall be:
(a) the continuation of the Outline under the sub-base to the trench side nearest that carriageway, hard shoulder or hard strip; and
(b) a vertical line being an extension of the trench side separating the contiguous fill from the trench filling.
In all cases of french drains the Earthworks Outline shall be the top of the trench filling filter material.

**Sewers, Drains, Piped Culverts and Ducts (excluding French Drains)**

**Units**

**3** The units of measurement for sewers, drains, piped culverts and ducts shall be:
 (i) sewers, drains, piped culverts, ducts............... linear metre.
 (ii) connections to existing sewers, drains, manholes, catchpits.................number.
 (iii) connections of permanently severed land drains to new drain.................number.

**Measurement**

**4** The measurement of sewers, drains, piped culverts and ducts shall be the summation of their individual lengths measured along the centre lines of the pipes between any of the following:
(a) the external faces of manholes, catchpits, drawpits, gullies or headwalls;
(b) the intersections of the centre lines at pipe junctions;
(c) the position of terminations shown on the Drawings;
(d) the point at which the depth to invert is 1·5 metres.

**5** The depth of sewers, drains, piped culverts and ducts shall be the vertical measurement between the invert and the following:

# Drainage

(a) where the invert level is below the Existing Ground Level—the Existing Ground Level except that where the Earthworks Outline is below the Existing Ground Level the measurement shall be taken to the Earthworks Outline;
(b) where the invert is at or above the Existing Ground Level—the Earthworks Outline;
(c) where a length of pipeline in or beneath areas of embankment or other areas of fill is not permitted to be constructed until a specified minimum level of cover is obtained then notwithstanding sub-paragraphs 5(a) and (b) above—to the specified minimum level of cover or the Earthworks Outline whichever is the lower.

**6** The average depth to invert shall be the calculated arithmetic mean of the measurements of depths taken at intervals of 10 metres along the pipelines starting from the outfall ends. For terminal lengths or pipelines less than 10 metres long the measurement of depth shall be taken at their ends.

**7** The measurement of ducts shall be for the complete construction irrespective of the number of ducts contained within any one trench.
Where more than one duct is laid in a trench then the number of ducts shall be stated in the item description.

**Itemisation**

**8** Separate items shall be provided for sewers, drains, piped culverts and ducts (excluding french drains) in accordance with Part II paragraphs 3 and 4 and the following:

| Group | Feature |
|---|---|
| I | 1 Connections to existing sewers or drains. |
| | 2 Connections to existing manholes or catchpits. |
| | 3 Connections of permanently severed land drains to new drain. |
| | 4 Sewers or drains. |
| | 5 Piped culverts. |
| | 6 Ducts. |
| II | 1 Different diameters. |
| III | 1 Depths to invert of 1·5 metres or less. |
| | 2 Depths to invert of over 1·5 metres the average and maximum depth to invert being stated to the nearest 25 mm. |
| IV | 1 Specified permitted alternative designs. |
| | 2 Particular designs at special locations indicated on the Drawings. |
| V | 1 Construction in trench. |

**Drainage** 30

|  |  |
|---|---|
|  | 2 Construction in heading.<br>3 Construction by jacking or thrust boring.<br>4 Suspended on discrete supports. |
| VI | 1 In side slopes of cuttings or embankments. |

Note: For each item which includes feature III 2 an associated item shall be provided for adjustment of the rate for each 25 mm of difference in excess of 150 mm where the average depth to invert calculated from site measurement varies from that stated in the Bill of Quantities.

**Connections to Existing Sewers or Drains**

**Item coverage**

**9** The items for connections to existing sewers or drains shall in accordance with the Preambles to Bill of Quantities General Directions include for:
(a) excavation of top soil (as Section 6 paragraph 12);
(b) excavation of suitable material (as Section 6 paragraph 13);
(c) excavation of unsuitable material (as Section 6 paragraph 14);
(d) locating and breaking into sewers and drains, including dealing with flow;
(e) backfilling with suitable material from any source and compaction;
(f) disposal (as Section 6 paragraph 25) of unsuitable material and surplus suitable material.

**Connections to Existing Manholes or Catchpits**

**Item coverage**

**10** The items for connections to existing manholes or catchpits shall in accordance with the Preambles to Bill of Quantities General Directions include for:
(a) breaking into manholes or catchpits, concrete benching and channel, dealing with flow and making good the benching, channel and walls;
(b) disposal (as Section 6 paragraph 25) of unsuitable material and surplus suitable material.

**Connections of Permanently Severed Land Drains to New Drain Item coverage**

**11** The items for connections of permanently severed land drains to new drain shall in accordance with the Preambles to Bill of Quantities General Directions include for:
(a) locating severed ends of land drains;
(b) excavation beyond width of main drain trench to facilitate connection, loading into transport, upholding the sides and keeping the earthworks free of water;
(c) junction on new drain;
(d) short length of drain and jointing to junction including laying and compacting pipe bedding;
(e) backfilling with suitable material from any source and compaction;
(f) disposal (as Section 6 paragraph 25) of unsuitable material and surplus suitable material;
(g) sealing off the intercepted downstream head.

Drainage 31

**Sewers, Drains, Piped Culverts and Ducts**

**Item coverage**

12 The items for sewers, drains, piped culverts and ducts shall in accordance with the Preambles to Bill of Quantities General Directions include for:
  (a) excavation of top soil (as Section 6 paragraph 12);
  (b) excavation of suitable material (as Section 6 paragraph 13);
  (c) excavation of unsuitable material (as Section 6 paragraph 14);
  (d) stripping turf and reserving for reuse, or if surplus to requirements, haulage and deposition in tips off Site provided by the Contractor;
  (e) access shafts to headings and their subsequent reinstatement;
  (f) thrust pits and thrust blocks for pipe jacking and their removal on completion;
  (g) cutting, laying, jointing and bedding pipes and fittings;
  (h) building in pipes to headwalls and outfall works;
  (i) hangers, stools and discrete supports;
  (j) reinstating or connecting into new drain severed land drains disturbed or damaged including sealing off downstream head;
  (k) laying and compacting pipe bedding, haunching and surround material;
  (l) formwork (as Section 14 paragraph 4);
  (m) backfilling with suitable material from any source and compaction;
  (n) disposal (as Section 6 paragraph 25) of unsuitable material and surplus suitable material;
  (o) movement joints to concrete bed and surrounds to flexible jointed pipes;
  (p) reinstatement of topsoil and turf;
  (q) cleaning;
  (r) recording, staking and labelling of junctions and terminations;
  (s) in the case of ducts fixing draw ropes, removable stoppers, marker blocks and posts.

**French Drains**

**Units**

13 The units of measurement for french drains shall be:
  (i) french drains.................linear metre.
  (ii) filter material contiguous with french drains......
       ...........cubic metre.
  (iii) connections to existing sewers, drains, manholes, catchpits.................number.

**Measurement**

14 The measurement of french drains shall be as for sewers, drains, piped culverts and ducts.

# Drainage

**Itemisation**

15 Separate items shall be provided for french drains in accordance with Part II paragraphs 3 and 4 and the following:

| Group | Feature |
|---|---|
| I | 1 Connections to existing sewers or drains.<br>2 Connections to existing manholes or catchpits.<br>3 French drains.<br>4 Filter material contiguous with french drains. |
| II | 1 Different diameters.<br>2 Different types of filter material. |
| III | 1 Depths to invert of 1·5 metres or less the average depth to invert being stated to the nearest 25 mm.<br>2 Depths to invert of over 1·5 metres the average and maximum depth to invert being stated to the nearest 25 mm. |
| IV | 1 Specified permitted alternative designs.<br>2 Particular designs at special locations indicated on the Drawings. |
| V | 1 In side slopes of cuttings or embankments. |

Note: For each item which includes feature III 1 or III 2 an associated item shall be provided for adjustment of the rate for each 25 mm of difference in excess of 150 mm where the average depth to invert calculated from site measurement varies from that stated in the Bill of Quantities.

**Connections to Existing Sewers or Drains**

**Item coverage**

16 *The items for connections to existing sewers or drains shall in accordance with the Preambles to Bill of Quantities General Directions include for:*
(a) *excavation of top soil (as Section 6 paragraph 12);*
(b) *excavation of suitable material (as Section 6 paragraph 13);*
(c) *excavation of unsuitable material (as Section 6 paragraph 14);*
(d) *locating and breaking into sewers and drains, including dealing with flow;*
(e) *backfilling with suitable material from any source and compaction;*
(f) *disposal (as Section 6 paragraph 25) of unsuitable material and surplus suitable material.*

**Connections to Existing Manholes or Catchpits**

17 *The items for connections to existing manholes or catchpits shall in accordance with the Preambles to Bill of Quantities General Directions include for:*

# Drainage

**Item coverage**

(a) breaking into manholes or catchpits, concrete benching and channel, dealing with flow, and making good the benching, channel and walls;

(b) disposal (as Section 6 paragraph 25) of unsuitable material and surplus suitable material.

**French Drains**

**Item coverage**

18 The items for french drains shall in accordance with the Preambles to Bill of Quantities General Directions include for:

(a) excavation of top soil (as Section 6 paragraph 12);

(b) excavation of suitable material (as Section 6 paragraph 13);

(c) excavation of unsuitable material (as Section 6 paragraph 14);

(d) stripping turf and reserving for reuse, or if surplus to requirements, haulage and deposition in tips off Site provided by the Contractor;

(e) disposal (as Section 6 paragraph 25) of unsuitable material and surplus suitable material;

(f) cutting, laying, jointing and bedding pipes and fittings;

(g) reinstating or connecting into new drain severed land drains disturbed or damaged including sealing off downstream head;

(h) laying and compacting pipe bedding, haunching and surround material;

(i) formwork (as Section 14 paragraph 4);

(j) filter material and compaction;

(k) reinstatement of topsoil and turf;

(l) cleaning;

(m) recording, staking and labelling of junctions and terminations.

**Filter Material Contiguous with French Drains**

**Item coverage**

19 The items for filter material contiguous with french drains shall in accordance with the Preambles to Bill of Quantities General Directions include for:

(a) compaction.

## Manholes, Catchpits, Drawpits and Gullies

**Units**

20 The units of measurement shall be:

(i) manholes, catchpits, drawpits, gullies ............ ............number.

**Measurement**

21 The measurement shall be of the complete manhole, catchpit, drawpit or gully.

22 Depths of manholes shall be the distance between the top surface of the cover and the invert of the main channel. Depths of catchpits and drawpits shall be the distance between the top surface of the cover or grating and the uppermost surface of the base slab.

Drainage

**Itemisation**

23 Separate items shall be provided for manholes, catchpits, drawpits and gullies in accordance with Part II paragraphs 3 and 4 and the following:

| Group | Feature |
|---|---|
| I | 1 Manholes.<br>2 Catchpits.<br>3 Drawpits.<br>4 Gullies. |
| II | 1 Specified permitted alternative designs.<br>2 Particular designs at special locations indicated on the Drawings. |
| III | 1 Depths to invert of 2 metres or less.<br>2 Depths to invert exceeding 2 metres but not exceeding 3 metres and thereafter in steps of 1 metre. |
| IV | 1 Different types of covers or gratings. |

**Manholes, Catchpits, Drawpits**

**Item coverage**

24 The items for manholes, catchpits and drawpits shall in accordance with the Preambles to Bill of Quantities General Directions include for:

(a) *excavation of top soil (as Section 6 paragraph 12);*
(b) *excavation of suitable material (as Section 6 paragraph 13);*
(c) *excavation of unsuitable material (as Section 6 paragraph 14);*
(d) *locating existing sewers and drains;*
(e) *construction of bases, walls, roof and cover slabs and shafts including in situ concrete (as Section 16 paragraph 4) surrounds and corbelling for cover;*
(f) *channels, benchings, building in pipe connections including fittings and pipe;*
(g) *connecting existing sewers and drains including dealing with flow;*
(h) *constructing vertical backdrop including concrete surround and all fittings;*
(i) *cleaning;*
(j) *step irons, safety chains, ladders, handholds and other fittings;*
(k) *covers, frames and bedding;*
(l) *cover keys;*
(m) *formwork (as Section 14 paragraph 4);*
(n) *reinforcement (as Section 15 paragraph 5);*
(o) *backfilling with in situ concrete (as Section 16 paragraph 4) or suitable material from any source and compaction;*

Drainage 35

(p) *disposal (as Section 6 paragraph 25) of unsuitable material and surplus suitable material.*

**Gullies**

**Item coverage**

25 *The items for gullies shall in accordance with the Preambles to Bill of Quantities General Directions include for:*
(a) *excavation of top soil (as Section 6 paragraph 12);*
(b) *excavation of suitable material (as Section 6 paragraph 13);*
(c) *excavation of unsuitable material (as Section 6 paragraph 14);*
(d) *gully chamber and fittings including in situ concrete (as Section 16 paragraph 4) bed and surround and jointing to pipes;*
(e) *gratings and frames including brickwork or concrete seating;*
(f) *formwork (as Section 14 paragraph 4);*
(g) *cleaning;*
(h) *backfilling with in situ concrete (as Section 16 paragraph 4) or suitable material from any source and compaction;*
(i) *disposal (as Section 6 paragraph 25) of unsuitable material and surplus suitable material.*

### Intercepting Ditches

**Units**

26 The units of measurement shall be:
   (i) excavation.................cubic metre.
   (ii) lining of ditches.................square metre.

**Measurement**

27 The quantities of excavation shall be the content of the voids formed by the removal of materials based on the net cross-section sufficient to contain any specified lining.

28 The measurement of lining shall be the face area of the work.

**Itemisation**

29 Separate items shall be provided for intercepting ditches in accordance with Part II paragraphs 3 and 4 and the following:

| Group | Feature |
|---|---|
| I | 1 Excavation of intercepting ditches. |
|   | 2 Lining of inverts to intercepting ditches. |
|   | 3 Lining of side slopes to intercepting ditches. |
| II | 1 Lining of different types. |
| III | 1 Linings of different thicknesses. |

Drainage

**Excavation of Intercepting Ditches**

**Item coverage**

30 The items for excavating intercepting ditches shall in accordance with the Preambles to Bill of Quantities General Directions include for:
(a) excavation of top soil (as Section 6 paragraph 12);
(b) excavation of suitable material (as Section 6 paragraph 13);
(c) excavation of unsuitable material (as Section 6 paragraph 14);
(d) stripping turf and reserving for reuse, or if surplus to requirements, haulage and deposition in tips off Site provided by the Contractor;
(e) disposal (as Section 6 paragraph 25) of unsuitable material and surplus suitable material;
(f) trimming of side slopes and inverts;
(g) locating and reinstating or connecting into new drain severed land drains disturbed or damaged including sealing off downstream head.

**Lining of Inverts and Side Slopes of Intercepting Ditches Item coverage**

31 The items for lining of inverts and side slopes of intercepting ditches shall in accordance with the Preambles to Bill of Quantities General Directions include for:
(a) dealing with the flow of water;
(b) levelling and compaction of bedding material;
(c) laying, setting, bedding, jointing, wedging, cutting and pointing;
(d) building in pipes;
(e) formwork (as Section 14 paragraph 4);
(f) reinforcement (as Section 15 paragraph 5);
(g) in situ concrete (as Section 16 paragraph 4).

### Headwalls and Outfall Works

**Units
Measurement
Itemisation**

32 Headwall and outfall works shall be measured in accordance with the units of measurement, measurement and itemisation set out in the relevant Sections hereof for Earthworks and Structures.

### Excavation in Rock and Reinforced Concrete

**Units**

33 The units of measurement shall be:
(i) excavation in rock, excavation in reinforced concrete................cubic metre. Measured extra over main construction.

**Measurement**

34 The measurement shall be the volume of the voids formed by the removal of the rock and reinforced concrete within the widths, lengths and depths required in the Contract for sewers, drains, piped culverts, ducts, french drains, manholes, catchpits, drawpits, gullies, headings, thrustpits and the like.

# Drainage

**Itemisation**     35 Separate items shall be provided for excavation in rock and reinforced concrete in accordance with Part II paragraphs 3 and 4 and the following:

| Group | Feature |
|---|---|
| I | 1 Excavation in rock.<br>2 Excavation in reinforced concrete. |

**Excavation in Rock and Reinforced Concrete**

**Item coverage**     36 The items for excavation in rock and reinforced concrete shall in accordance with the Preambles to Bill of Quantities General Directions include for:
(a) preliminary site trials of blasting;
(b) loosening or breaking up by blasting or other means of unexcavated material before or in the process of excavation;
(c) cutting through reinforcement;
(d) breaking down of material necessary to comply with the requirements of fill;
(e) trimming the bottom and sides of excavation and clearing away loose rock;
(f) over break and making good.

## Reinstatement of Pavement

**Units**     37 The units of measurement shall be:
    (i) reinstatement of pavement over pits and trenches .................. linear metre. Measured extra over main construction.

**Measurement**     38 Measurement shall be along the centre line of the drain under and across manholes.

**Itemisation**     39 Separate items shall be provided for reinstatement of pavement in accordance with Part II paragraphs 3 and 4 and the following:

| Group | Feature |
|---|---|
| I | 1 Reinstatement of pavement. |
| II | 1 Different types of reinstatement. |
| III | 1 Different diameters. |

**Reinstatement of Pavement**

**Item coverage**     40 The items for reinstatement of pavement shall in accordance with the Preambles to Bill of Quantities General Directions include for:
(a) grading, measuring, mixing, spreading and compacting

Drainage 38

> materials, tack coats, forming and sealing joints, finishing, curing and protecting;
> (b) formwork (as Section 14 paragraph 4);
> (c) reinforcement (as Section 15 paragraph 5);
> (d) additional pavement around manholes and the like;
> (e) making good new work up to existing including cutting back to expose reinforcement for bonding.

## Soft Spots and Other Voids

**Units**

41 The units of measurement shall be:
   (i) soft spots, other voids ................... cubic metre.

**Measurement**

42 The measurement shall be the volume of the soft spots and other voids within the widths and lengths required in the Contract for sewers, drains, piped culverts, ducts, french drains, manholes, catchpits, drawpits, gullies, headings, thrustpits and the like measured below the thinnest permitted bed in any one group for trenches and from the underside of the base slab of manholes and the like.

**Itemisation**

43 Separate items shall be provided for soft spots and other voids in accordance with Part II paragraphs 3 and 4 and the following:

| Group | Feature |
|---|---|
| I | 1 Excavation of soft spots. <br> 2 Filling of soft spots and other voids. |
| II | 1 Bottoms of trenches, manholes, catchpits, drawpits and gullies. |
| III | 1 Different types of fill. |

**Excavation of Soft Spots**

**Item coverage**

44 The items for excavation of soft spots shall in accordance with the Preambles to Bill of Quantities General Directions include for:
(a) loading into transport;
(b) haulage and deposition in tips off Site provided by the Contractor;
(c) keeping the earthworks free of water;
(d) upholding the sides;
(e) multiple handling of excavated material.

**Filling of Soft Spots and Other Voids**

**Item coverage**

45 The items for filling of soft spots and other voids shall in accordance with the Preambles to Bill of Quantities General Directions include for:
(a) keeping the earthworks free of water;

# Drainage

(*b*) *material, deposition and compaction;*
(*c*) *formwork* (*as Section 14 paragraph 4*).

## Supports left in Excavation

**Units**     46 The units of measurement shall be:
       (i) supports left in excavation ............... square metre.

**Measurement**     47 Measurement shall be the area of face ordered by the Engineer to be left with supports in position.

**Itemisation**     48 Separate items shall be provided for supports left in excavation in accordance with Part II paragraphs 3 and 4 and the following:

| Group | Feature |
|---|---|
| I | 1 Timber supports.<br>2 Steel sheeting supports. |
| II | 1 Construction in trench.<br>2 Construction in pits.<br>3 Construction in heading. |

**Supports left in Excavation**

**Item coverage**     49 *The items for supports left in excavation shall in accordance with the Preambles to Bill of Quantities General Directions include for:*
(*a*) *the value of the sheeting, struts and waling materials left in excavation;*
(*b*) *working around struts, walings etc during backfilling.*

## Drainage and Ducts in Bridges, Viaducts and Other Structures

**Units**     50 The units of measurement shall be:
       (i) drainage and ducts in substructure and superstructure of bridges and viaducts, drainage and ducts in other structures ................. item.

**Itemisation**     51 Separate items shall be provided for drainage and ducts in substructure and superstructure of bridges and viaducts and drainage and ducts in other structures in accordance with Part II paragraphs 3 and 4 and the following:

| Group | Feature |
|---|---|
| I | 1 Drainage.<br>2 Ducts. |

Drainage 40

| II | 1 Substructure—End Supports.<br>2 Substructure—Intermediate Supports.<br>3 Superstructure.<br>4 Other structure. |

**Drainage and Ducts in Substructure and Superstructure of Bridges and Viaducts and Drainage and Ducts in Other Structures**
**Item coverage**

**52** The items for drainage and ducts in substructure and superstructure of bridges and viaducts and drainage and ducts in other structures shall in accordance with the Preambles to Bill of Quantities General Directions include for:

(a) sewers, drains, piped culverts and ducts (as this Section paragraph 12);
(b) manholes, catchpits and drawpits (as this Section paragraph 24);
(c) gullies (as this Section paragraph 25);
(d) fixing to or building into the structure pipework, gullies, downpipes, fittings and the like including brackets, hangers and straps;
(e) collecting drains and drainage outfalls and making connection to other drainage systems;
(f) forming channels;
(g) making good protective treatment;
(h) free drainage layer including compaction and supports.

### Filling to Pipe Bays on Bridges

**Units**

**53** The units of measurement shall be:
   (i) filling to pipe bays on bridges.................. cubic metre.

**Measurement**

**54** Measurement shall be the volume of the void shown on the Drawings or ordered by the Engineer to be filled except that no deduction shall be made for sewers, drains and ducts.

**Itemisation**

**55** Separate items shall be provided for filling to pipe bays on bridges in accordance with Part II paragraphs 3 and 4 and the following:

| Group | Feature |
|---|---|
| I | 1 Filling to pipe bays on bridges. |
| II | 1 Different types of filling material. |

**Filling to Pipe Bays on Bridges**

**56** The items for filling to pipe bays on bridges shall in accordance with the Preambles to Bill of Quantities General Directions include for:

## Drainage 41

**Item coverage**

(a) *deposition;*
(b) *allowance for bulking and shrinkage;*
(c) *complying with any restrictions on the placing and compacting of materials;*
(d) *compaction around sewers, drains, ducts, services, fittings and the like;*
(e) *taking precautions to avoid damage to the structure, sewers, drains, ducts, services, fittings and the like.*

# Section 6: Earthworks

1 For the purpose of this Section it shall be assumed that one cubic metre of material excavated forms one cubic metre of compacted fill and no allowance shall be made in the measurement for bulking and shrinkage.

2 No account shall be taken of excavated material arising from the Works measured in accordance with Sections 1 to 5 and 11 to 13 hereof.

3 Where:
(a) settlement occurs subsequent to the Earthworks Outline having been reached and the compaction of the embankment has been completed in accordance with the Contract; or
(b) settlement of or penetration into the ground beneath the embankment occurs;
then the additional deposition, fill and compaction required shall, subject to the proviso below, be measured immediately prior to the laying of the sub-base.
Provided that the first 75 mm of settlement or penetration into the ground beneath the embankment shall not be measured.

**Definitions**

4 For the purpose of this Section Sub-soil Level is defined as the level of the ground after the removal of top soil required by and measured under the Contract.

5 The Earthworks Outline is defined as the finished earthworks levels and dimensions required by the Contract for the construction of the carriageway, hard shoulder, hard strip, paved area, sub-base material under verge, central reserve and side slope, footway, topsoiling and turfing.

6 Where the filling between a french drain and an adjacent carriageway, hard shoulder or hard strip is of filter material (contiguous filter material) the Earthworks Outline shall be:
(a) the continuation of the Outline under the sub-base to the trench side nearest that carriageway, hard shoulder or hard strip; and
(b) a vertical line being an extension of the trench side separating the contiguous fill from the trench filling.
In all cases of french drains the Earthworks Outline shall be the top of the trench filling filter material.

7 Where the bottom of the structural foundation for an earth

# Earthworks

retaining structure is below the Existing Ground Level, the Earthworks Outline shall be the permanently exposed face of the structure below the Existing Ground Level.

## Excavation

**Units**

8 The units of measurement shall be:
  (i) excavation.................cubic metre.

**Measurement for Cutting and Excavation in Bulk in the Open**

9 The measurement for cuttings and excavations in bulk in the open shall be:
(a) for cuttings—the volume of the material removed (other than top soil) to form the Earthworks Outline and voids formed by the excavation of material below that Outline; or
(b) under embankments and other areas of fill—the volume of the void formed by the excavation of material (other than top soil) beneath the embankment and other areas of fill;
but excluding in both cases the excavation of material measured as soft spots under paragraph 39 of this Section.

**Measurement for Structural Foundations**

10 The measurement for structural foundations shall be the volume of the void formed to accommodate the structural foundation. The calculation shall be based on the horizontal area of the bottom of the foundation with the depth being measured from the bottom of the foundation to the following:
(a) where the bottom of the structural foundation is below the Existing Ground Level—the Existing Ground Level or the Sub-soil Level whichever is applicable. Provided that where the Earthworks Outline is below either the Existing Ground Level or the Sub-soil Level, whichever is appropriate, then the depth shall be measured to the Earthworks Outline;
(b) where the bottom of the structural foundation is at or above the Existing Ground Level—the Earthworks Outline.
The classification of stage depth for the excavation of structural foundations shall be the maximum depth in any one excavation obtained in accordance with this paragraph.

**Itemisation**

11 Separate items shall be provided for excavation in accordance with Part II paragraphs 3 and 4 and the following:

| Group | Feature |
|---|---|
| I | 1 Top soil. |
|   | 2 Suitable material except rock. |
|   | 3 Unsuitable material. |
|   | 4 Rock. |
|   | 5 Reinforced concrete. |

## Earthworks 44

| | |
|---|---|
| II | 1 Cuttings and excavations in bulk in the open.<br>2 Structural foundations.<br>3 New watercourses.<br>4 Enlarged watercourses.<br>5 Clearing abandoned watercourses. |
| III | 1 Structural foundations 0 metres to 3 metres in depth.<br>2 Structural foundations 0 metres to 6 metres in depth and so on in steps of 3 metres. |

Note: Top soil shall not be separately identified by any Group II or III feature.

**Excavation of Top Soil**

**Item coverage**

12 *The items for excavation of top soil shall in accordance with the Preambles to Bill of Quantities General Directions include for:*
(a) *loading into transport;*
(b) *haulage and deposition in temporary stockpiles including the provision of sites for stockpiles;*
(c) *where surplus to requirements haulage and deposition in tips off Site provided by the Contractor;*
(d) *multiple handling of excavated material;*
(e) *keeping the earthworks free of water.*

**Excavation of Suitable Material**

**Item coverage**

13 *The items for excavation of suitable material shall in accordance with the Preambles to Bill of Quantities General Directions include for:*
(a) *loading into transport;*
(b) *protection of the sub-grade with suitable material and its subsequent removal;*
(c) *replacing suitable material rendered unsuitable;*
(d) *loosening or breaking up material before or in the process of excavation;*
(e) *multiple handling of excavated material;*
(f) *keeping the earthworks free of water;*
(g) *upholding the sides;*
(h) *waiting for frozen suitable material to thaw;*
(i) *selection and separation of materials;*
(j) *breaking down of material necessary to comply with the requirements of fill;*
(k) *forming and trimming side slopes, benching and berms, or in the case of structural foundations, trimming the bottom and sides of excavation;*
(l) *taking precautions to avoid damage to property, structures, sewers, drains and services;*
(m) *excavating around and between piles;*
(n) *additional excavation the Contractor may require for working space, timbering, formwork or other temporary*

# Earthworks

works and its subsequent backfilling with approved material and compaction;
(o) in the case of watercourses and culverts, grading beds and working in and dealing with the flow of water;
(p) haulage, deposition and compaction in temporary stockpiles including the provision of sites for stockpiles.

**Excavation of Unsuitable Material**

**Item coverage**

**14** The items for excavation of unsuitable material shall in accordance with the Preambles to Bill of Quantities General Directions include for:
(a) loading into transport;
(b) loosening or breaking up materials before or in the process of excavation;
(c) multiple handling of excavated material;
(d) keeping the earthworks free of water;
(e) upholding the sides;
(f) selection and separation of materials;
(g) forming and trimming side slopes, benchings and berms, or in the case of structural foundations, trimming the bottom and sides of excavation;
(h) taking precautions to avoid damage to property, structures, sewers, drains and services;
(i) excavating around and between piles;
(j) additional excavation the Contractor may require for working space, timbering, formwork or other temporary works and its subsequent backfilling with approved material and compaction;
(k) in the case of watercourses and culverts, grading beds and working in and dealing with the flow of water.

**Excavation in Rock and Reinforced Concrete**

**Item coverage**

**15** The items for excavation in rock and reinforced concrete shall in accordance with the Preambles to Bill of Quantities General Directions include for:
(a) preliminary site trials of blasting;
(b) cutting through reinforcement;
(c) loading into transport;
(d) protection of the sub-grade with suitable material and its subsequent removal;
(e) replacing suitable material rendered unsuitable;
(f) loosening or breaking up by blasting or other means of material before or in the process of excavation;
(g) selection and separation of materials;
(h) multiple handling of excavated material;
(i) keeping the earthworks free of water;
(j) upholding the sides;
(k) breaking down of material necessary to comply with the requirements of fill;
(l) over break and making good;
(m) forming and trimming side slopes, benching and berms, or in the case of structural foundations, trimming the

# Earthworks 46

        *bottom and sides of excavation and clearing away loose rock;*
- (n) *taking precautions to avoid damage to property, structures, sewers, drains and services;*
- (o) *excavating around and between piles;*
- (p) *additional excavation the Contractor may require for working space, timbering, formwork or other temporary works and its subsequent backfilling with approved material and compaction;*
- (q) *in the case of watercourses and culverts, grading beds and working in and dealing with the flow of water;*
- (r) *haulage, deposition and compaction in temporary stockpiles including the provision of sites for stockpiles.*

### Deposition of Fill

**Units**

**16** The units of measurement shall be:
    (i)  deposition of fill..................cubic metre.

**Measurement of Deposition of Fill**

**17** Deposition of fill shall be measured separately for each of the Group III features. The measurement shall be the volume of compacted fill, calculated in accordance with paragraphs 33 to 35 inclusive, less the volume of the imported fill calculated in accordance with paragraphs 27 and 28.

**18** Deposition of rock fill shall only be separately measured where rock fill as such is specifically shown on the Drawings or ordered by the Engineer.

**Itemisation**

**19** Separate items shall be provided for deposition of fill in accordance with Part II paragraphs 3 and 4 and the following:

| Group | Feature |
|---|---|
| I | 1 Deposition. |
| II | 1 Suitable material.<br>2 Rock fill. |
| III | 1 In embankments and other areas of fill other than adjacent to structures.<br>2 On sub-base material under verges, central reserves and side slopes.<br>3 Into soft areas.<br>4 Around structural foundations.<br>5 Adjacent to structures excluding around foundations.<br>6 On bridges (under footways, verges and central reserves). |

Earthworks 47

**Deposition of Fill**

**Item coverage**

20 The items for deposition of fill shall in accordance with the Preambles to Bill of Quantities General Directions include for:
(a) haulage;
(b) selection of material of stated types and layering or depositing in locations indicated on Drawings;
(c) the mechanical or chemical treatment of soil as the Contractor may require to facilitate the use of particular plant;
(d) complying with requirements for the equalisation of earth pressures or the sequence or rate of deposition;
(e) depositing fill to slope away from vertical drainage layers and providing temporary drainage to prevent surface water from entering such drainage layers;
(f) multiple handling of excavated material;
(g) keeping earthworks free of water;
(h) waiting for frozen material to thaw;
(i) trimming and shaping to required levels and contours;
(j) taking precautions to avoid damage to property, structures, sewers, drains and services;
(k) protection of the sub-grade with suitable material and its subsequent removal;
(l) replacing suitable material rendered unsuitable;
(m) deposition of the first 75 mm of fill resulting from the settlement of or penetration into the underlying material beneath embankments.

**Disposal of Material**

**Units**

21 The units of measurement shall be:
(i) disposal of material.................cubic metre.

**Measurement of Disposal of Suitable Material (Including Rock and Reinforced Concrete)**

22 The measurement of disposal of suitable material (including rock and reinforced concrete) shall be the volume of that suitable material excavated from within the Site measured in this Section less the volume of compacted fill, calculated in accordance with paragraphs 33 to 35 inclusive, after deduction from the latter of the volume of imported fill calculated in accordance with paragraphs 27 and 28.

**Measurement of Disposal of Unsuitable Material**

23 The measurement of disposal of unsuitable material shall be the volume of unsuitable material excavated and measured in accordance with paragraphs 9 and 10 together with the unsuitable material arising from the excavation of watercourses.

**Itemisation**

24 Separate items shall be provided for disposal of material in accordance with Part II paragraphs 3 and 4 and the following:

| Group | Feature |
|---|---|
| I | 1 Disposal. |

Earthworks 48

| | | |
|---|---|---|
| II | 1 | Suitable material. |
| | 2 | Unsuitable material. |

| | | |
|---|---|---|
| III | 1 | In tips off Site. |

**Disposal of Suitable Material and Unsuitable Material**

**Item coverage**

25 *The items for disposal of suitable material and unsuitable material shall in accordance with the Preambles to Bill of Quantities General Directions include for:*
(a) *haulage and deposition in tips off Site provided by the Contractor;*
(b) *multiple handling of excavated material;*
(c) *in the case of disposal of suitable material allowing for the deposition in lieu of disposal of the fill required for the first 75 mm of settlement or penetration into the underlying material beneath embankments.*

**Imported Fill**

**Units**

26 The units of measurement shall be:
    (i) imported fill.................cubic metre.

**Measurement of Imported Fill**

27 The measurement of imported suitable material fill shall be the volumes of the compacted fill, calculated in accordance with paragraphs 33 to 35 inclusive, less the volumes of:
(a) the suitable material excavated from within the Site measured in this Section;
(b) other stated types of imported fill.

28 The measurement of other stated types of imported fill shall be to the outlines shown on the Drawings or ordered by the Engineer.

29 Imported rock fill shall only be separately measured where rock fill as such is specifically shown on the Drawings or ordered by the Engineer.

**Itemisation**

30 Separate items shall be provided for imported fill in accordance with Part II paragraphs 3 and 4 and the following:

| Group | Feature | |
|---|---|---|
| I | 1 | Imported suitable material fill. |
| | 2 | Other stated types of imported fill. |
| II | 1 | Deposited in embankments and other areas of fill other than adjacent to structures. |
| | 2 | Deposited on sub-base material under verges, central reserves and side slopes. |
| | 3 | Deposited into soft areas. |
| | 4 | Deposited around structural foundations. |

# Earthworks

5 Deposited adjacent to structures excluding around foundations.
6 Deposited on bridges (under footways, verges and central reserves).

**Imported Fill**

**Item coverage**

31 The items for imported fill shall in accordance with the Preambles to Bill of Quantities General Directions include for:
(a) *fill provided by the Contractor from sources outside the Site;*
(b) *selection of material of stated types and layering or depositing in locations indicated on Drawings;*
(c) *complying with requirements for the equalisation of earth pressures or the sequence or rate of filling;*
(d) *depositing fill to slope away from vertical drainage layers and providing temporary drainage to prevent surface water from entering such drainage layers;*
(e) *keeping the earthworks free of water;*
(f) *trimming and shaping to required levels and contours;*
(g) *taking precautions to avoid damage to property, structures, sewers, drains and services;*
(h) *suitable material required for the protection of the sub-grade and its subsequent removal;*
(i) *replacing suitable material rendered unsuitable;*
(j) *imported fill for the first 75 mm resulting from the settlement of or penetration into the underlying material beneath embankments.*

## Compaction of Fill

**Units**

32 The units of measurement shall be:
  (i) compaction ................. cubic metre.

**Measurement**

33 The measurement of compaction of fill in embankments and other areas of fill other than adjacent to structures shall be the volume of the embankment formed or void filled up to the Earthworks Outline from the Existing Ground Level or Sub-soil Level, whichever is the lower, together with the filling of any void the excavation of which has been measured in accordance with paragraph 9 of this Section.

34 The measurement of compaction of fill around structural foundations shall be the volume of the void excavated, measured in accordance with paragraph 10, less the volume of the structural foundation and structure within that void.

35 The measurement of compaction of fill:
(a) On sub-base material under verges, central reserves and side slopes;
(b) Adjacent to structures excluding around foundations;
(c) On bridges (under footways, verges and central reserves);

## Earthworks 50

shall be the volume of the void filled to the outline shown on the Drawing or ordered by the Engineer.

**36** Compaction of rock fill shall only be separately measured where rock fill as such is specifically shown on the Drawings or ordered by the Engineer.

**Itemisation**

**37** Separate items shall be provided for compaction of fill in accordance with Part II paragraphs 3 and 4 and the following:

| Group | Feature |
|---|---|
| I | 1 Compaction of fill material. <br> 2 Compaction of rock fill. |
| II | 1 In embankments and other areas of fill other than adjacent to structures. <br> 2 On sub-base material under verges, central reserves and side slopes. <br> 3 Into soft areas. <br> 4 Around structural foundations. <br> 5 Adjacent to structures excluding around foundations. <br> 6 On bridges (under footways, verges and central reserves). |

**Compaction of Fill**

**Item coverage**

**38** The items for compaction of fill shall in accordance with the Preambles to Bill of Quantities General Directions include for:

(a) *spreading and levelling;*
(b) *spreading, levelling and compaction of suitable material for the protection of the sub-grade and its subsequent removal;*
(c) *keeping the earthworks free of water;*
(d) *operations necessary to produce an acceptable ground surface where the method of deposition and compaction of material into soft areas or below water is adopted;*
(e) *blinding of side slopes;*
(f) *forming and trimming of side slopes, benchings and berms;*
(g) *taking precautions to avoid damage to property, structures, sewers, drains and services;*
(h) *complying with requirements for the equalisation of earth pressures or the sequence or rate of filling;*
(i) *trials by the Contractor to demonstrate compaction methods different from those stated in the Contract except that materials compacted by such trials and accepted as part of the Works will be included in the measurement of the Works;*
(j) *compaction of the first 75 mm of fill resulting from the*

# Earthworks

*settlement of or penetration into the underlying material beneath embankment.*

## Soft Spots and Other Voids

**Units**

39 The units of measurement shall be:
  (i) soft spots, other voids................cubic metre.

**Measurement**

40 Measurement shall be made separately from the main excavation only where the volume:
(a) below structural foundations or in rock in side slopes is less than 1 cubic metre; and
(b) elsewhere is less than 25 cubic metres.

**Itemisation**

41 Separate items shall be provided for soft spots and other voids in accordance with Part II paragraphs 3 and 4 and the following:

| Group | Feature |
|---|---|
| I | 1 Excavation of soft spots.<br>2 Filling of soft spots and other voids. |
| II | 1 Below cuttings or under embankments.<br>2 In rock in side slopes.<br>3 Below structural foundations. |
| III | 1 Different types of fill. |

**Excavation of Soft Spots**

**Item coverage**

42 *The items for excavation of soft spots shall in accordance with the Preambles to Bill of Quantities General Directions include for:*
*(a) loading into transport;*
*(b) haulage and deposition in tips off Site provided by the Contractor;*
*(c) keeping the earthworks free of water;*
*(d) upholding the sides;*
*(e) multiple handling of excavated material.*

**Filling of Soft Spots and Other Voids**

**Item coverage**

43 *The items for filling of soft spots and other voids shall in accordance with the Preambles to Bill of Quantities General Directions include for:*
*(a) keeping the earthworks free of water;*
*(b) material, deposition and compaction;*
*(c) formwork (as Section 14 paragraph 4).*

## Supports left in Excavation

**Units**

44 The units of measurement shall be:
  (i) supports left in excavation................square metre.

Earthworks

| | |
|---|---|
| **Measurement** | **45** Measurement shall be the area of face ordered by the Engineer to be left with supports in position. |
| **Itemisation** | **46** Separate items shall be provided for supports left in excavation in accordance with Part II paragraphs 3 and 4 and the following: |

| Group | Feature |
|---|---|
| I | 1 Timber supports.<br>2 Steel sheeting supports. |

| | |
|---|---|
| **Supports left in Excavation** | **47** *The items for supports left in excavation shall in accordance with the Preambles to Bill of Quantities General Directions include for:* |
| **Item coverage** | (a) *value of the sheeting, struts and walings left in excavation;*<br>(b) *working around struts, walings and the like during backfilling.* |

### Soiling

| | |
|---|---|
| **Units** | **48** The units of measurement shall be:<br>(i) soiling................square metre. |
| **Itemisation** | **49** Separate items shall be provided for soiling in accordance with Part II paragraphs 3 and 4 and the following: |

| Group | Feature |
|---|---|
| I | 1 Soiling of different thicknesses. |
| II | 1 Surfaces sloping at 10° or less to the horizontal.<br>2 Surfaces sloping at more than 10° to the horizontal. |

| | |
|---|---|
| **Soiling** | **50** *The items for soiling shall in accordance with the Preambles to Bill of Quantities General Directions include for:* |
| **Item coverage** | (a) *the removal of debris from the surface of earthworks to be soiled;*<br>(b) *excavation from stockpile, loading into transport, haulage, deposition, spreading, levelling and compaction of top soil on verges, central reserves and side slopes of cuttings and embankments and, where described in the Contract, other areas of fill.* |

### Grassing

| | |
|---|---|
| **Units** | **51** The units of measurement shall be:<br>(i) grass seeding, turfing...............square metre. |

Earthworks

**Itemisation**

52 Separate items shall be provided for grassing in accordance with Part II paragraphs 3 and 4 and the following:

| Group | Feature |
|---|---|
| I | 1 Grass seeding.<br>2 Turfing.<br>3 'Hydraulic Mulch' grass seeding. |
| II | 1 Surfaces sloping at 10° or less to the horizontal.<br>2 Surfaces sloping at more than 10° to the horizontal. |

**Grass Seeding and Turfing**

**Item coverage**

53 The items for grass seeding and turfing shall in accordance with the Preambles to Bill of Quantities General Directions include for:
(a) freeing surfaces of areas to be grassed or turfed from stones and other debris and reducing the soil to a tilth immediately prior to grassing;
(b) fertilising including additional plant nutrients;
(c) mowing and clearance of grass cuttings;
(d) pegging and wiring of turves.

## Completion of Formation

**Units**

54 The units of measurement shall be:
   (i) completion of formation.................square metre.

**Measurement**

55 The measurement shall be the area calculated from the dimensions shown on the Drawings of all surfaces below carriageway, hard shoulder, hard strip, paved area, sub-base material under verge, central reserve and side slope, and footway. Completion of formation on rock fill shall only be separately measured where rock fill as such is specifically shown on the Drawings or ordered by the Engineer.

**Itemisation**

56 Separate items shall be provided for completion of formation in accordance with Part II paragraphs 3 and 4 and the following:

| Group | Feature |
|---|---|
| I | 1 Formation of suitable material.<br>2 Formation of rock fill.<br>3 Formation of rock cuttings. |
| II | 1 Regulated with lean concrete.<br>2 Regulated with sub-base material. |

# Earthworks

**Completion of Formation**

**Item coverage**

57 *The items for completion of formation shall in accordance with the Preambles to Bill of Quantities General Directions include for:*
 (a) *preliminary cleaning, trimming, regulating and rolling of the surface;*
 (b) *taking measures to protect the sub-grade from deterioration due to the ingress of water and the use of constructional plant;*
 (c) *blinding of rock fill.*

## Lining of Watercourses

**Units**

58 The units of measurement shall be:
 (i) lining of watercourses....................square metre.

**Measurement**

59 Measurement of lining shall be the face area of the work.

**Itemisation**

60 Separate items shall be provided for lining of watercourses in accordance with Part II paragraphs 3 and 4 and the following:

| Group | Feature |
|---|---|
| I | 1 Lining of new watercourse. |
|   | 2 Lining of enlarged watercourse. |
| II | 1 To inverts. |
|   | 2 To side slopes. |
| III | 1 Different types. |
| IV | 1 Different thicknesses. |

**Lining of Watercourses**

**Item coverage**

61 *The items for lining of watercourses shall in accordance with the Preambles to Bill of Quantities General Directions include for:*
 (a) *dealing with the flow of water;*
 (b) *levelling and compaction of bedding material;*
 (c) *laying, setting, bedding, jointing, wedging, cutting and pointing;*
 (d) *building in pipes;*
 (e) *formwork (as Section 14 paragraph 4);*
 (f) *reinforcement (as Section 15 paragraph 5);*
 (g) *in situ concrete (as Section 16 paragraph 4).*

## Clearing of Existing Ditches

**Units**

62 The units of measurement shall be:
 (i) clearing of existing ditches..................linear metre.

## Earthworks

**Measurement**

63 The measurement of clearing of existing ditches shall be the length along the centre of the ditch.

**Itemisation**

64 Separate items shall be provided for clearing of existing ditches in accordance with Part II paragraphs 3 and 4 and the following:

| Group | Feature |
|---|---|
| I | 1 Clearing of existing ditches. |
| II | 1 At differing locations. |

**Clearing of Existing Ditches**

**Item coverage**

65 *The items for clearing of existing ditches shall in accordance with the Preambles to Bill of Quantities General Directions include for:*
(a) *excavation and loading into transport;*
(b) *haulage and deposition of excavated materials in tips off Site provided by the Contractor;*
(c) *clearing and removing debris and weeds to tips off Site provided by the Contractor;*
(d) *working in and dealing with the existing flow of water;*
(e) *trimming side slopes and grading bottoms;*
(f) *maintaining existing outfalls.*

# Section 7: Roadworks Overall Requirements

Separate items shall not be provided for roadworks overall requirements described in the series No. 700 clauses of the Specification.

# Section 8: Sub-base and Roadbase

**Units**

1 The units of measurement shall be:
  (i) sub-base.................cubic metre.
  (ii) roadbase.................square metre.
  (iii) bituminous regulating course (conversion factor per cubic metre per tonne to be stated)................ tonne.
  (iv) cement based regulating course.................... cubic metre.

**Measurement**

2 The measurement of sub-base shall be the volume of that material measured to the outlines shown on the Drawings or ordered by the Engineer.

3 The measurement of roadbase whether laid in one or more layers shall be calculated using the outlines shown on the Drawings or ordered by the Engineer, including any projection into the verge or central reserve, and shall be described as in carriageway or hard shoulder as appropriate.
Where carriageway roadbase projects into the hard shoulder and is of a differing thickness to the roadbase in that hard shoulder, then the projection shall be measured with and described as roadbase in carriageway.
The roadbase in hard strip shall be measured with and described as roadbase in carriageway.

4 The measurement of bituminous regulating course shall be calculated from the superficial area laid to the outline shown on the Drawing or ordered by the Engineer multiplied by the average thickness and the conversion factor quoted in the Bill of Quantities irrespective of the type of bituminous regulating material supplied by the Contractor.

5 No deduction shall be made for openings of 1 square metre or less.

**Itemisation**

6 Separate items shall be provided for sub-base, roadbase and regulating course in accordance with Part II paragraphs 3 and 4 and the following:

| Group | Feature |
|---|---|
| I | 1 Each group or type of sub-base. |
|   | 2 Each group or type of roadbase. |

## Sub-base and Roadbase                                                        58

|     |   |                                          |
|-----|---|------------------------------------------|
|     | 3 | Each group or type of regulating course. |
| II  | 1 | Roadbase in carriageway.                 |
|     | 2 | Roadbase in hard shoulder.               |
| III | 1 | Roadbases of different thicknesses.      |

**Sub-base Roadbase and Regulating Course**

**Item coverage**

**7** *The items for sub-base, roadbase and regulating course shall in accordance with the Preambles to Bill of Quantities General Directions include for:*
(a) *preliminary trials;*
(b) *trials by the Contractor to demonstrate different compaction methods from those stated in the Contract;*
(c) *materials and attendance for sampling and testing carried out by the Engineer;*
(d) *protection of mixed material in transit and while awaiting tipping;*
(e) *grading, measuring, mixing and depositing materials;*
(f) *spreading and compaction of deposited material in layers;*
(g) *making joints;*
(h) *curing cement treated materials;*
(i) *edge support;*
(j) *maintenance of surface;*
(k) *taking measures to protect the sub-grade, sub-base and roadbase from deterioration due to the ingress of water and the use of constructional plant;*
(l) *taking measures to increase the thickness and/or strength of the sub-base to protect the sub-base and sub-grade from damage due to the Contractors method of construction and choice of constructional plant;*
(m) *shaping to cambers, falls and crowns.*

# Section 9: Flexible Surfacing

**Units**

1 The units of measurement shall be:
   (i) flexible surfacing, slurry sealing, surface dressing, bituminous spray..................square metre.
   (ii) regulating course (conversion factor per cubic metre per tonne to be stated)..................tonne.

**Measurement**

2 The measurement of flexible surfacing shall be calculated using the width of the top surface of the surfacing consisting of basecourse, wearing course, slurry sealing, surface dressing or bituminous spray either singly or in combination as described in the Contract.

3 The measurement of flexible surfacing shall include any projection of the flexible surfacing into the verge or central reserve and shall be described as flexible surfacing in carriageway or hard shoulder as appropriate.

Where carriageway flexible surfacing projects into the hard shoulder and is of differing thickness to the flexible surfacing in that hard shoulder, then the projection shall be measured with and described as flexible surfacing in carriageway.

Flexible surfacing in hard strip shall be measured with and described as flexible surfacing in carriageway.

4 Slurry sealing, surface dressing and bituminous spray forming an integral part of any specified group or type of flexible surfacing shall not be separately measured. Such surface treatments shall only be measured when shown on the Drawings or ordered by the Engineer separately from or in addition to any specified flexible surfacing.

5 The measurement of regulating course shall be calculated from the superficial area laid to the outline shown on the Drawing or ordered by the Engineer multiplied by the average thickness and the conversion factor quoted in the Bill of Quantities irrespective of the type of regulating material supplied by the Contractor.

6 No deduction shall be made for openings of 1 square metre or less.

7 Trial lengths of slurry sealing will be measured only if they are accepted as part of the Permanent Works.

**Itemisation**

8 Separate items shall be provided for flexible surfacing in accordance with Part II paragraphs 3 and 4 and the following:

## Flexible Surfacing

| Group | Feature |
|---|---|
| I | 1 Each group or type of flexible surfacing.<br>2 Each group or type of regulating course.<br>3 Slurry sealing.<br>4 Each type of surface dressing.<br>5 Each type of bituminous spray. |
| II | 1 Carriageway.<br>2 Hard shoulder. |
| III | 1 Different thicknesses.<br>2 Different rates of spread. |

**Flexible Surfacing and Regulating Course**

**Item coverage**

9 The items for flexible surfacing and regulating course shall in accordance with Preambles to Bill of Quantities General Directions include for:
(a) materials and attendance for sampling and testing carried out by the Engineer;
(b) grading, measuring, mixing and depositing materials;
(c) spreading and compaction of deposited material in layers including chippings, surface dressing, bituminous spray and slurry sealing;
(d) making joints;
(e) edge support;
(f) preliminary trial lengths of slurry sealing not accepted as part of the Permanent Works;
(g) site control tests of slurry sealing;
(h) protection of kerbs, masking and unmasking of drainage channels, manhole and catchpit covers, gully gratings, expansion joints, road studs and road markings and obtaining clean lines;
(i) making good up to abutting surfaces including cleaning and painting with asphaltic cement;
(j) shaping to cambers, falls and crowns;
(k) measures required for aftercare and opening the road to traffic.

**Slurry Sealing Surface Dressing and Bituminous Spray Item coverage**

10 The items for slurry sealing, surface dressing and bituminous spray shall in accordance with the Preambles to Bill of Quantities General Directions include for:
(a) materials and attendance for sampling and testing carried out by the Engineer;
(b) grading, measuring, mixing and depositing materials;
(c) spreading and rolling deposited materials;
(d) making joints;
(e) preliminary trial lengths of slurry sealing not accepted as part of the Permanent Works;
(f) site control tests of slurry sealing;

## Flexible Surfacing

(g) *protection of kerbs, masking and unmasking of drainage channels, manhole and catchpit covers, gully gratings, expansion joints, road studs and road markings and obtaining clean lines;*
(h) *measures required for aftercare and opening the road to traffic.*

# Section 10: Concrete Pavement

**Units**

1 The units of measurement shall be:
  (i) concrete carriageway, concrete hard shoulder......
  ............square metre.

**Measurement**

2 The measurement of concrete carriageway and hard shoulder shall be calculated using the width of the top surface of the slab. Trial lengths shall be measured only if they are accepted as part of the Permanent Works.

3 The measurement of concrete carriageway and hard shoulder shall include any projection of the concrete into the verge or central reserve and shall be described as concrete carriageway or hard shoulder as appropriate.
Where concrete carriageway projects into the hard shoulder and is of differing thickness to the concrete in that hard shoulder, then the projection shall be measured with and described as concrete carriageway.
Concrete hard strip shall be measured with and described as concrete carriageway.

4 No deduction shall be made for openings of 1 square metre or less.

**Itemisation**

5 Separate items shall be provided for concrete pavement in accordance with Part II paragraphs 3 and 4 and the following:

| Group | Feature |
|---|---|
| I | 1 Concrete carriageway.<br>2 Concrete hard shoulder. |
| II | 1 Specified permitted alternative designs.<br>2 Reinforced.<br>3 Unreinforced. |
| III | 1 Different thicknesses. |

**Concrete Carriageway and Concrete Hard Shoulder Item coverage**

6 The items for concrete carriageway and concrete hard shoulder shall in accordance with the Preambles to Bill of Quantities General Directions include for:
  (a) preliminary trial lengths not accepted as part of the Permanent Works;
  (b) awaiting Engineer's approval of trial lengths;

## Concrete Pavement

(c) trial mixes;
(d) materials and attendance for sampling and testing carried out by the Engineer;
(e) protection of mixed material in transit and while awaiting tipping;
(f) waterproof membrane underlay;
(g) longitudinal, expansion, contraction, warping and construction joint assemblies, including joint filler and crack inducers, tie and dowel bars, dowel bar caps and bond breaking compound;
(h) formwork (as Section 14 paragraph 4);
(i) mixing, laying, spreading, compaction and finishing;
(j) air entrainment;
(k) reinforcement (as Section 15 paragraph 5);
(l) forming or sawing grooves, cleaning, caulking, temporary and permanent sealing of joints;
(m) surface texturing including surface texturing in hardened state of accepted trial lengths, curing and protection;
(n) forming recesses, openings and bays;
(o) shaping to falls.

# Section 11: Kerbs and Footways

**Kerbing, Channelling and Edging**

**Units**

1 The units of measurement shall be:
　(i) kerbing, channelling, edging ............... linear metre.

**Measurement**

2 The measurement of kerbing, channelling and edging shall be the lengths of the work. No deduction shall be made for gaps of 1 linear metre or less.

**Itemisation**

3 Separate items shall be provided for kerbing, channelling and edging in accordance with Part II paragraphs 3 and 4 and the following:

| Group | Feature |
|---|---|
| I | 1 Kerbing.<br>2 Channelling.<br>3 Edging. |
| II | 1 Specified permitted alternative materials and designs.<br>2 Particular materials and designs. |
| III | 1 Straight or curved over 12 metres radius.<br>2 Curves of 12 metres radius or less. |

**Kerbing, Channelling and Edging**

**Item coverage**

4 *The items for kerbing, channelling and edging shall in accordance with the Preambles to Bill of Quantities General Directions include for:*
(a) *trial mixes;*
(b) *materials and attendance for sampling and testing carried out by the Engineer;*
(c) *excavation and loading into transport, upholding the sides and keeping the earthworks free of water;*
(d) *formwork (as Section 14 paragraph 4);*
(e) *reinforcement (as Section 15 paragraph 5);*
(f) *in situ concrete (as Section 16 paragraph 4) in foundation and backings;*
(g) *bedding and jointing;*
(h) *mixing materials and extruding kerbs;*
(i) *making, filling and sealing joints;*
(j) *keying of surfaces and tack coats;*
(k) *surface finishing, curing and protecting;*
(l) *building in gully gratings and frames;*

# Kerbs and Footways

(m) *tie bars;*
(n) *drainage holes or pipes through concrete;*
(o) *quadrants and other special units;*
(p) *disposal of surplus material.*

### Excavation in Rock and Reinforced Concrete

**Units**

5 The units of measurement shall be:
(i) excavation in rock, excavation in reinforced concrete ................... cubic metre. Measured extra over Kerbing, Channelling and Edging.

**Measurement**

6 The measurement shall be the volume of the voids formed by the removal of the rock and reinforced concrete.
For this measurement, the dimensions of trenches shall be taken as the dimensions shown on the Drawings or ordered by the Engineer.

**Itemisation**

7 Separate items shall be provided for excavation in rock and reinforced concrete in accordance with Part II paragraphs 3 and 4 and the following:

| Group | Feature |
|---|---|
| I | 1 Excavation in rock. |
|   | 2 Excavation in reinforced concrete. |

**Excavation in Rock and Reinforced Concrete**

**Item coverage**

8 *The items for excavation in rock and reinforced concrete shall in accordance with the Preambles to Bill of Quantities General Directions include for:*
(a) *preliminary site trials of blasting;*
(b) *loosening or breaking up by blasting or other means of unexcavated material before or in the process of excavation;*
(c) *cutting through reinforcement;*
(d) *trimming the bottom and sides of excavation and clearing away loose rock;*
(e) *over break and making good.*

### Footways

**Units**

9 The units of measurement shall be:
(i) concrete paved footways..................... square metre.

**Measurement**

10 The measurement of concrete paved footways shall be calculated using the width of the top surface of the footway.

11 No deduction shall be made for openings of 1 square metre or less.

## Kerbs and Footways 66

**Itemisation**

**12** Separate items shall be provided for footways in accordance with Part II paragraphs 3 and 4 and the following:

| Group | Feature |
|---|---|
| I | 1 Different types of concrete paved footways. |
| II | 1 Different sizes. |
| III | 1 Different thicknesses. |
| IV | 1 Different thickness of sub-base. |

**Concrete Paved Footways**

**Item coverage**

**13** *The items for concrete paved footways shall in accordance with the Preambles to Bill of Quantities General Directions include for:*
*(a) sub-base material including laying and compaction;*
*(b) formwork (as Section 14 paragraph 4);*
*(c) laying to falls;*
*(d) bedding and jointing;*
*(e) straight, circular and radial cutting and fitting.*

# Section 12: Traffic Signs and Road Markings

**Traffic Signs**

**Unit**  1 The unit of measurement shall be:
(i) traffic signs.................number.

**Measurement**  2 The measurement of traffic signs shall be of the complete installation.

**Itemisation**  3 Separate items shall be provided for traffic signs in accordance with Part II paragraphs 3 and 4 and the following:

| Group | Feature |
|---|---|
| I | 1 Standard traffic signs.<br>2 Non-standard traffic signs. |
| II | 1 Different types of signs. |
| III | 1 Externally illuminated signs.<br>2 Internally illuminated signs. |
| IV | 1 Reflectorised signs.<br>2 Unreflectorised signs. |
| V | 1 Different sizes. |
| VI | 1 Different supports. |

**Traffic Signs**

**Item coverage**  4 The items for traffic signs shall in accordance with the Preambles to Bill of Quantities General Directions include for:
(a) *excavation and loading into transport, upholding the sides and keeping the earthworks free of water;*
(b) *backfilling and compaction;*
(c) *supports;*
(d) *formwork (as Section 14 paragraph 4);*
(e) *reinforcement (as Section 15 paragraph 5);*
(f) *in situ concrete (as Section 16 paragraph 4);*
(g) *ducts to bases;*
(h) *electrical equipment, except electricity supply cabling;*
(i) *reinstatement of surfaces;*
(j) *temporary screening of signs until they become operative;*
(k) *disposal of surplus material;*
(l) *supply of Drawings showing detail design.*

## Traffic Signs and Road Markings

### Road Markings

**Units**

5 The units of measurement shall be:
 (i) solid areas..................square metre.
 (ii) lines...................linear metre.
 (iii) triangles, circles with enclosing arrows, arrows, kerb markings...................number. (The Diagram number from the Traffic Signs Regulations and General Directions to be stated.)
 (iv) letters and numerals...................number.

**Measurement**

6 Solid areas shall only be measured for the solid infilling between converging lines, the enclosing lines themselves shall be measured as lines.

7 Markings other than those measured under sub-paragraphs 5(i), (iii) and (iv) above shall be measured as lines and shall be grouped together according to width.
In the case of intermittent lines the measurement shall be of the marks only but the length of the mark and gap shall be stated. Double lines shall be measured as two single lines. Diagonal lines between double lines and short transverse lines at the ends of any longitudinal lines shall be measured with the lines of which they form part.
Ancillary lines shall include lines forming hatched areas, chevrons, zigzag lines and boxed areas (including their enclosing lines). In the case of hatched areas and chevrons the enclosing lines shall be measured with the longitudinal line of which they form part. The measurement of zigzag lines shall include any transverse or longitudinal lines at their ends.

8 The measurement of circles with enclosing arrows (mini roundabouts) shall be for the complete marking, the external diameter of the circle being stated. Distinction shall be made for all other arrows between straight, curved, turning or double headed.

9 Kerb markings shall be measured as a single item irrespective of the number of lines forming any one-marking.

10 Each letter or numeral shall be separately measured with all letters or numerals grouped together according to height.

**Itemisation**

11 Separate items shall be provided for road markings in accordance with Part II paragraphs 3 and 4 and the following:

| Group | Feature |
|---|---|
| I | 1 Solid areas. |
|   | 2 Continuous lines. |
|   | 3 Intermittent lines. |
|   | 4 Ancillary lines. |
|   | 5 Triangles. |

## Traffic Signs and Road Markings

|     |                                              |
| --- | -------------------------------------------- |
|     | 6 Circle with enclosing arrows.              |
|     | 7 Arrows.                                    |
|     | 8 Kerb markings.                             |
|     | 9 Letters and Numerals.                      |
| II  | 1 Different materials.                       |
| III | 1 Different width of lines.                  |
|     | 2 Different diameter of circles.             |
|     | 3 Different lengths of arrows.               |
|     | 4 Different lengths of kerb markings.        |
|     | 5 Different heights of letters and numerals. |
| IV  | 1 Different length of mark and gap for intermittent lines. |
|     | 2 Different diagram numbers for arrows and kerb markings. |
| V   | 1 Different types of arrows.                 |

**Solid Areas, Lines, Triangles, Circles with Enclosing Arrows, Kerb Markings and Letters and Numerals Item coverage**

12 *The items for solid areas, lines, triangles, circles with enclosing arrows, kerb markings and letters and numerals shall in accordance with Preambles to Bill of Quantities General Directions include for:*

(a) *cleaning and drying surfaces;*
(b) *application of the marking materials including the incorporation of specified reflecting medium;*
(c) *tack coat;*
(d) *apostrophes in the case of letters and numerals;*
(e) *kerb markings down the face of kerbs.*

### Road Studs

**Unit**

13 The unit of measurement shall be:
   (i) road studs.................number.

**Itemisation**

14 Separate items shall be provided for road studs in accordance with part II paragraphs 3 and 4 and the following:

| Group | Feature             |
| ----- | ------------------- |
| I     | 1 Road Studs.       |
| II    | 1 Different types.  |
| III   | 1 Different shapes. |
| IV    | 1 Different sizes.  |

Traffic Signs and Road Markings 70

**Road Studs**

**Item coverage**

15 *The items for road studs shall in accordance with the Preambles to Bill of Quantities General Directions include for:*
(a) *forming of holes;*
(b) *adhesives;*
(c) *reinstatement of surfaces;*
(d) *disposal of surplus material.*

**Marker Posts**

**Unit**

16 The unit of measurement shall be:
(i) marker posts.................number.

**Itemisation**

17 Separate items shall be provided for marker posts in accordance with Part II paragraphs 3 and 4 and the following:

| Group | Feature |
|---|---|
| I | 1 Marker posts. |
| II | 1 Different types. |

**Marker Posts**

**Item coverage**

18 *The items for marker posts shall in accordance with the Preambles to Bill of Quantities General Directions include for:*
(a) *painting;*
(b) *numerals, symbols and reflectorised strips or discs including adhesive;*
(c) *driving or excavation and loading into transport, upholding the sides and keeping the earthworks free of water;*
(d) *backfilling and compaction;*
(e) *sockets;*
(f) *fixings and fittings including galvanising;*
(g) *preservation of timber;*
(h) *disposal of surplus material.*

**Excavation in Rock and Reinforced Concrete**

**Units**

19 The units of measurement shall be:
(i) excavation in rock, excavation in reinforced concrete.................cubic metre. Measured extra over Traffic Signs.

**Measurement**

20 The measurement shall be the volume of the voids formed by the removal of the rock and reinforced concrete.
The dimensions of post holes and foundations shall be taken as the dimensions shown on the **Drawings** or ordered by the **Engineer**.

**Itemisation**

21 Separate items shall be provided for excavation in rock and reinforced concrete in accordance with Part II paragraphs 3 and 4 and the following:

Traffic Signs and Road Markings 71

| Group | Feature |
|---|---|
| I | 1 Excavation in rock.<br>2 Excavation in reinforced concrete. |

**Excavation in Rock and Reinforced Concrete**

**Item coverage**

22 *The items for excavation in rock and reinforced concrete shall in accordance with the Preambles to Bill of Quantities General Directions include for:*
(a) *preliminary site trials of blasting;*
(b) *loosening or breaking up by blasting or other means of unexcavated material before or in the process of excavation;*
(c) *cutting through reinforcement;*
(d) *breakdown of material to comply with the requirements of fill;*
(e) *trimming the bottom and sides of excavation and clearing away loose rock;*
(f) *over break and making good.*

# Section 13: Piling for Structures

**Piling Plant**

**Units**

1 The units of measurement shall be:
(i) establishment of piling plant.................item.
(ii) moving piling plant.................number.

**Measurement**

2 The measurement of moving piling plant shall correspond to the number of piles. Any additional moves to suit the Contractors method of working shall not be measured.

3 Moving of piling plant shall not be measured for steel sheet piling.

**Itemisation**

4 Separate items shall be provided for piling plant in accordance with Part II paragraphs 3 and 4 and the following:

| Group | Feature |
|---|---|
| I | 1 Establishment of piling plant. <br> 2 Moving piling plant. |
| II | 1 Piling plant for different types of pile. |
| III | 1 Trial piling as a separate operation in advance of the main piling. <br> 2 Main piling. |

**Establishment of Piling Plant**

**Item coverage**

5 *The items for the establishment of piling plant shall in accordance with the Preambles to Bill of Quantities General Directions include for:*
(a) *bringing plant and equipment to the site of the bridge viaduct or other structure;*
(b) *erecting and setting up plant and equipment including site preparation, levelling and access ramps;*
(c) *dismantling and removing plant and equipment from site on completion;*

**Moving Piling Plant**

**Item coverage**

6 *The items for moving piling plant shall in accordance with the Preambles to Bill of Quantities General Directions include for:*
(a) *moving and setting up plant and equipment at each pile position including site preparation or levelling.*

Piling 73

### Precast Concrete Piles

**Units**

7 The units of measurement shall be:
(i) precast concrete piles, driving precast concrete piles, lengthening precast concrete piles, driving lengthened precast concrete piles.................linear metre.
(ii) stripping precast concrete pile heads................ number.

**Measurement**

8 The measurement of precast concrete piles shall be the lengths shown on the Drawings or ordered by the Engineer.

9 The measurement of driving precast concrete piles shall be the length of each pile measured along the axis from the toe of the shoe or tapered point to:
(a) the existing ground level or the level of the underside of the pile cap or ground beam (ignoring any blinding layer) whichever is the lower. Provided that where a particular level is specified from which driving shall commence, then the measurement shall be to that specified level; or
(b) the site joint of piles to be lengthened provided that the site joint after completion of the driving is below the level determined in accordance with the preceding sub-paragraph.

10 The measurement of lengthening precast concrete piles shall be the lengths of the added concrete ordered by the Engineer.

11 The measurement of driving lengthened precast concrete piles shall be the length from the site joint to the level determined in accordance with sub-paragraph 9(a) above.

12 The length classification for those items listed in sub-paragraph 7(i) above shall be based on the appropriate measured lengths determined in accordance with paragraphs 8 to 11 above inclusive.

**Itemisation**

13 Separate items shall be provided for precast concrete piles in accordance with Part II paragraphs 3 and 4 and the following:

| Group | Feature |
|---|---|
| I | 1 Precast concrete piles.<br>2 Driving precast concrete piles.<br>3 Lengthening precast concrete piles.<br>4 Driving lengthened precast concrete piles.<br>5 Stripping precast concrete pile heads. |
| II | 1 Vertical.<br>2 Raking. |

Piling 74

| III | 1 Different types. |
|---|---|
| IV | 1 Different materials. |
| V | 1 Different cross section. |
| VI | 1 Piles not exceeding 5 metres in length.<br>2 Piles exceeding 5 metres in length but not exceeding 10 metres and so on in steps of 5 metres. |
| VII | 1 Trial piling as a separate operation in advance of the main piling.<br>2 Main piling. |

**Precast Concrete Piles**

**Item coverage**

14 *The items for precast concrete piles shall in accordance with the Preambles to Bill of Quantities General Directions include for:*
(a) *formwork (as Section 14 paragraph 4);*
(b) *reinforcement (as Section 15 paragraph 5);*
(c) *in situ concrete and precast members, segmental units and the like (as Section 16 paragraphs 4 and 9);*
(d) *pile shoes, tapered points, chamfers, prefabricated joints and joint fitments.*

**Driving Precast Concrete Piles**

**Item coverage**

15 *The items for driving precast concrete piles shall in accordance with the Preambles to Bill of Quantities General Directions include for:*
(a) *pre-boring or where permitted by the Engineer, jetting;*
(b) *handling, pitching and driving to a given set or level;*
(c) *taking observations and compiling the record of each pile and supplying one copy to the Engineer;*
(d) *moving plant and equipment back and redriving risen piles.*

**Lengthening Precast Concrete Piles**

**Item coverage**

16 *The items for lengthening precast concrete piles shall in accordance with the Preambles to Bill of Quantities General Directions include for:*
(a) *stripping concrete to provide reinforcement bond lengths;*
(b) *forming connection with the old work including lapping, tying or welding reinforcement;*
(c) *formwork (as Section 14 paragraph 4);*
(d) *reinforcement (as Section 15 paragraph 5);*
(e) *in situ concrete (as Section 16 paragraph 4);*
(f) *lost time, moving plant and equipment, standing time and disruption caused by the process of lengthening including waiting for concrete to achieve a specified strength or age.*

**Driving Lengthened Precast Concrete Piles**

17 *The items for driving, lengthened precast concrete piles shall in accordance with the Preambles to Bill of Quantities General Directions include for:*

## Piling

**Item coverage**
(a) *driving to a given set or level;*
(b) *taking observations and compiling the record of each pile and supplying one copy to the Engineer;*
(c) *moving plant and equipment back and driving lengthened pile;*
(d) *moving plant and equipment back and redriving risen piles.*

**Stripping Precast Concrete Pile Heads**

**18** The items for stripping precast concrete pile heads shall in accordance with the Preambles to Bill of Quantities General Directions include for:

**Item coverage**
(a) *cutting off, removing and disposing of surplus;*
(b) *stripping concrete from the head to the required level and exposing the reinforcement required for bonding into the substructure;*
(c) *bending and tying the projecting reinforcement to bond into the substructure.*

### Cast-in-Place Piles

**Units**

**19** The units of measurement shall be:
    (i) pile shafts, empty bores.............linear metre.
    (ii) enlarged bases................number.
    (iii) steel bar reinforcement.................tonne.

**Measurement**

**20** The measurement of pile shafts shall be the length of the pile measured along the axis from the toe of the shoe or the bottom of the excavation, including any enlarged base, whichever is appropriate to the specified level of the concrete pile head.

Where an empty bore is specified above a pile shaft, the length classification of the pile shaft shall be based on the overall bored or driven depth including the empty bore.

**21** Empty bores shall only be measured where a particular level is specified from which boring or driving shall commence and shall be the length of empty bore or drive measured from the specified level of the concrete pile head, to that particular commencing level.

**22** The mass of plain bar reinforcement to cast-in-place piles shall be calculated on the basis that the nominal density of steel is 0·00785 kilogrammes per square millimetre of cross sectional area per linear metre; the mass of deformed bar reinforcement shall be calculated as the nominal rolling mass of the reinforcement.

The weight of the reinforcement shall include the reinforcement specified to be cast into the pile required to bond into the substructure.

**Itemisation**

**23** Separate items shall be provided for cast-in-place piles in accordance with Part II paragraphs 3 and 4 and the following:

Piling

| Group | Feature |
|---|---|
| I | 1 Pile shafts.<br>2 Empty bores.<br>3 Enlarged bases. |
| II | 1 Vertical.<br>2 Raking. |
| III | 1 Different types. |
| IV | 1 Different materials. |
| V | 1 Different cross section. |
| VI | 1 Pile shafts not exceeding 5 metres in length.<br>2 Pile shafts exceeding 5 metres in length but not exceeding 10 metres and so on in steps of 5 metres. |
| VII | 1 Trial piling as a separate operation in advance of the main piling.<br>2 Main piling. |

**Pile Shafts**

**Item coverage**

24 *The items for pile shafts shall in accordance with the Preambles to Bill of Quantities General Directions include for:*
(a) *pre-boring or where permitted by the Engineer, jetting;*
(b) *casing or lining;*
(c) *boring or augering to a given level and removing and disposing of surplus excavated material from the bore, or alternatively, driving to a given set or level with or without pile shoe;*
(d) *removing temporary casing or lining;*
(e) *providing apparatus including where appropriate protection to personnel entering the empty bore for the inspection of pile excavation or inside of pile casing, taking observations and maintaining the boring or driving record of each pile and supplying one copy to the Engineer;*
(f) *forming the pile including in situ concrete and precast members, segmental units and the like (as Section 16 paragraphs 4 and 9);*
(g) *cutting back concrete to form bond and or trimming to required finished level;*
(h) *taking measures required because of the presence of water in the bore or drive;*
(i) *taking undisturbed soil samples from the bore at any level;*
(j) *temporary filling around the top of piles and subsequent removal.*

**Empty Bore**

25 *The items for empty bore shall in accordance with the*

# Piling

**Item coverage**

*Preambles to Bill of Quantities General Directions include for:*
(a) pre-boring or where permitted by the Engineer, jetting;
(b) temporary casing or lining and removing;
(c) boring or augering to a given level and removing and disposing of surplus excavated material from the bore, or alternatively, driving to a given set or level, with or without pile shoe;
(d) providing apparatus including where appropriate protection to personnel entering the empty bore for the inspection of pile excavation or inside of pile casing, taking observations and maintaining the boring or driving record of each pile and supplying one copy to the Engineer;
(e) taking measures required because of the presence of water in the bore or drive;
(f) taking undisturbed soil samples from the bore at any level;
(g) temporary filling of empty bore and subsequent removal.

**Enlarged Bases**

**Item coverage**

**26** *The item for enlarged bases shall in accordance with the Preambles to Bill of Quantities General Directions include for:*
(a) under-reaming and removing and disposing of material excavated from the bore;
(b) placing and compacting or driving additional concrete or other material required to fill or form the enlarged base;
(c) taking measures required because of the presence of water.

### Reinforcement for Cast-in-Place Piles

**Itemisation**

**27** Separate items shall be provided for reinforcement for cast-in-place piles in accordance with Part II paragraphs 3 and 4 and the following:

| Group | Feature |
|---|---|
| I | 1 Bar reinforcement of nominal size 16 millimetres and under.<br>2 Bar reinforcement of nominal size 20 millimetres and over.<br>3 Helical reinforcement. |
| II | 1 Mild steel.<br>2 High yield steel.<br>3 Stainless steel. |
| III | 1 Bars of 12 metres length or less.<br>2 Bars exceeding 12 metres in length but not exceeding 13·5 metres and so on in steps of 1·5 metres. |

Piling 78

**Reinforcement**

**Item coverage**

28 *The items for reinforcement shall in accordance with the Preambles to Bill of Quantities General Directions include for:*
*(a) cleaning, cutting and bending;*
*(b) binding with wire or other material;*
*(c) supports and spacers;*
*(d) welding;*
*(e) bending projecting reinforcement to bond into the substructure.*

### Steel Sheet Piles

**Units**

29 The units of measurement shall be:
    (i)   steel sheet piling, driving steel sheet piling, lengthening pieces to steel sheet piling, driving lengthened steel sheet piling............square metre.
    (ii)  corner, junction, special steel sheet piles........
           ...........linear metre.
           Measured extra over steel sheet piling.
    (iii) lengthening pieces to corner, junction, special steel sheet piles................linear metre.
           Measured eytra over lengthening pieces to steel sheet piling.
    (iv) driving corner, junction, special steel sheet piles ................linear metre.
           Measured extra over driving steel sheet piling.
    (v)  driving lengthened corner, junction, special steel sheet piles................linear metre.
           Measured extra over driving lengthened steel sheet piling.
    (vi) welding on lengthening pieces............... linear metre.
    (vii) cutting or burning off surplus.............linear metre.
    (viii) walings, ties.................tonne.

**Measurement**

30 The measurement of steel sheet piling shall be the plane (not developed) horizontal length along the centre line of the piling multiplied by the length shown on the Drawing or ordered by the Engineer.

31 The measurement for driving steel sheet piling shall be the plane (not developed) horizontal length along the centre line of the piling multiplied by the depth from the toe to:
  (a) the existing ground level. Provided that where a particular level is specified from which driving shall commence, then the measurement shall be to that specified level; or
  (b) the site joint of piles to be lengthened provided that the site joint after completion of the driving, is below the level determined in accordance with the preceding sub-paragraph.

## Piling

**32** The measurement of lengthening pieces to steel sheet piling shall be the plane (not developed) horizontal length along the centre line of the piling multiplied by the additional length ordered by the Engineer.

**33** The measurement for driving lengthened steel sheet piling shall be the plane (not developed) horizontal length along the centre line of the piling multiplied by the depth from the site joint to the level determined in accordance sub-paragraph 31(a) above.

**34** The length classification for those items listed in sub-paragraphs 29(i) to (v) above shall be based on the appropriate measured lengths determined in accordance with paragraphs 30 to 33 above inclusive.

**35** The measurement of welding on lengthening pieces and cutting or burning off surplus shall be the plane (not developed) horizontal length along the centre line of the piling.

**Itemisation**

**36** Separate items shall be provided for steel sheet piles in accordance with Part II paragraphs 3 and 4 and the following:

| Group | Feature |
|---|---|
| I | 1 Steel sheet piling. |
| | 2 Driving steel sheet piling. |
| | 3 Corner, junction, special steel sheet piles. |
| | 4 Driving corner, junction, special steel sheet piles. |
| | 5 Lengthening pieces to steel sheet piling. |
| | 6 Driving lengthened steel sheet piling. |
| | 7 Lengthening pieces to corner, junction, special steel sheet piles. |
| | 8 Driving lengthened corner, junction, special steel sheet piles. |
| | 9 Welding on lengthening pieces. |
| | 10 Cutting or burning off surplus. |
| | 11 Walings. |
| | 12 Ties. |
| II | 1 Different types. |
| III | 1 Piles not exceeding 5 metres in length. |
| | 2 Piles exceeding 5 metres in length but not exceeding 10 metres and so on in steps of 5 metres. |
| IV | 1 In main construction. |
| | 2 In anchorages. |

| | |
|---|---|
| **Steel Sheet Piling, Corner, Junction, Special Steel Sheet Piles and Lengthening Pieces to Steel Sheet Piling, Corner, Junction, Special Steel Sheet Piles** <br> **Item coverage** | **37** The items for steel sheet piling, corner, junction, special steel sheet piles and lengthening pieces to steel sheet piling and corner, junction, special steel sheet piles shall be in accordance with the Preambles to Bill of Quantities General Directions include for: <br><br> (a) fabrication, lifting holes, tapered points and reference markings; <br> (b) protective system applied at place of manufacture. |
| **Driving Steel Sheet Piling, Corner, Junction, Special Steel Sheet Piles and Driving Lengthened Steel Sheet Piling, Corner, Junction, Special Steel Sheet Piles** <br> **Item coverage** | **38** The items for driving steel sheet piling, corner, junction, special steel sheet piles and driving lengthened steel sheet piling, corner, junction, special steel sheet piles shall in accordance with the Preambles to Bill of Quantities General Directions include for: <br><br> (a) handling, pitching and driving to a given set, level or penetration; <br> (b) moving and setting up plant and equipment at each pile position including site preparation or levelling; <br> (c) taking observations and compiling the record of the piling and supplying one copy to the Engineer; <br> (d) moving plant and equipment back and redriving risen piles. |
| **Welding on Lengthening Pieces** <br><br> **Item coverage** | **39** The items for welding on lengthening pieces shall in accordance with the Preambles to Bill of Quantities General Directions include for: <br> (a) stripping protective system and preparing the head of the driven pile to receive additional length; <br> (b) full penetration butt weld between driven pile and additional length; <br> (c) cleaning the affected area and applying protective system; <br> (d) lost time, moving plant and equipment, standing time and disruption caused by the process of lengthening; <br> (e) inspection and testing of welds. |
| **Cutting or Burning Off Surplus** <br><br> **Item coverage** | **40** The item for cutting or burning off surplus shall in accordance with the Preambles to Bill of Quantities General Directions include for: <br> (a) cutting or burning off to profile, removing and disposing of surplus; <br> (b) cleaning the affected area and applying protective system. |
| **Walings** <br><br> **Item coverage** | **41** The items for walings shall in accordance with the Preambles to Bill of Quantities General Directions include for: <br> (a) fabrication, handling and fixing including necessary excavation, back filling, compaction, disposal of surplus |

Piling 81

excavated material, upholding the sides and keeping earthworks free of water;
(b) marking off, cutting and drilling;
(c) nuts, bolts, washers, plates, welding and the like;
(d) protective system applied at the place of manufacture.

**Ties**

**Item coverage**

42 The items for ties shall in accordance with the Preambles to Bill of Quantities General Directions include for:
(a) fabrication, handling and fixing including necessary excavation, backfilling, compaction, disposal of surplus excavated material, upholding the sides and keeping earthworks free of water;
(b) nuts, bolts, washers, bearing plates, couplings, turnbuckles, welding and the like;
(c) marking off, cutting and threading;
(d) marking and drilling piles;
(e) protective system applied at the place of manufacture.

**Steel Bearing Piles**

**Units**

43 The units of measurement shall be:
   (i) steel bearing piles, driving steel bearing piles, lengthening pieces for steel bearing piles, driving lengthened steel bearing piles.................. linear metre.
   (ii) welding on lengthening pieces.................. number.
   (iii) cutting or burning off surplus.................... number.

**Measurement**

44 The measurement of steel bearing piles shall be the length shown on the Drawings or ordered by the Engineer.

45 The measurement of driving steel bearing piles shall be the length of the pile measured along the axis from the base to:
(a) the existing ground level or the level of the underside of the pile cap or ground beam (ignoring any blinding layer) whichever is the lower. Provided that where a particular level is specified from which driving shall commence, then the measurement shall be to that specified level; or
(b) the site joint of piles to be lengthened provided that the site joint after completion of the driving is below the level determined in accordance with the preceding sub paragraph.

46 The measurement of lengthening pieces for steel bearing piles shall be the additional length ordered by the Engineer.

47 The measurement of driving lengthened steel bearing piles shall be the length from the site joint to the level determined in accordance with sub-paragraph 45(a) above.

# Piling

**48** The length classification for those items listed in sub paragraph 43(i) above shall be based on the appropriate measured lengths determined in accordance with paragraphs 44 to 47 above inclusive.

**Itemisation**

**49** Separate items shall be provided for steel bearing piles in accordance with Part II, paragraphs 3 and 4 and the following:

| Group | Feature |
|---|---|
| I | 1 Steel bearing piles.<br>2 Driving steel bearing piles.<br>3 Lengthening pieces for steel bearing piles.<br>4 Driving lengthened steel bearing piles.<br>5 Welding on lengthening pieces.<br>6 Cutting or burning off surplus. |
| II | 1 Vertical.<br>2 Raking. |
| III | 1 Different types. |
| IV | 1 Piles not exceeding 5 metres in length.<br>2 Piles exceeding 5 metres in length but not exceeding 10 metres and so on in steps of 5 metres. |
| V | 1 Trial piling as a separate operation in advance of the main piling.<br>2 Main piling. |

**Steel Bearing Piles and Lengthening Pieces for Steel Bearing Piles Item coverage**

**50** The items for steel bearing piles and lengthening pieces for steel bearing piles shall in accordance with the Preambles to Bill of Quantities General Directions include for:
(a) fabrication, stiffening, slinging holes and reference markings;
(b) protective system applied at place of manufacture.

**Driving Steel Bearing Piles and Driving Lengthened Steel Bearing Piles Item coverage**

**51** The items for driving steel bearing piles and driving lengthened steel bearing piles shall in accordance with the Preambles to Bill of Quantities General Directions include for:

(a) pre-boring or where permitted by the Engineer, jetting;
(b) handling, pitching and driving to a given set, level or penetration;
(c) taking observations and compiling the record of each pile and supplying one copy to the Engineer;
(d) moving plant and equipment back and redriving risen piles.

# Piling

**Welding on Lengthening Pieces**

**Item coverage**

**52** The items for welding on lengthening pieces shall in accordance with the Preambles to Bill of Quantities General Directions include for:
(a) stripping protective system and preparing the head of the driven pile to receive additional length;
(b) full penetration butt weld between driven pile and additional length;
(c) cleaning the affected area and applying protective system;
(d) lost time, moving plant and equipment, standing time and disruption caused by the process of lengthening;
(e) inspection and testing of welds.

**Cutting or Burning Off Surplus**

**Item coverage**

**53** The items for cutting or burning off surplus shall in accordance with the Preambles to Bill of Quantities General Directions include for:
(a) cutting or burning off to profile, removing and disposing of surplus;
(b) cleaning the affected area and applying protective system.

# Section 14: Formwork for Structures

**Formwork**

**Units**

1 The units of measurement shall be:
  (i) formwork..................square metre.
  (ii) void formers.................linear metre.

**Measurement**

2 The measurement shall be the area of formwork which is in contact with the finished concrete but measured over the face of openings of 1 square metre or less and features described in (c) below.
Formwork shall not be measured:
(a) to construction joints (whether shown or not on the Drawings) skewbacks, stunt ends, steppings, bonding chases and the like;
(b) to holes, ducts, pockets, sockets, mortices and the like, not exceeding 0·15 cubic metres each in volume;
(c) to individual fillets, chamfers, splays, drips, rebates, recesses, grooves and the like of 100 mm total girth or less when measured overall the faces in contact with the concrete;
(d) to edge of blinding concrete 75 mm or less in thickness;
(e) to upper surfaces of concrete inclined at an angle of less than 15° to the horizontal.
Where concrete, other than blinding concrete 75 mm or less in thickness, is placed in structural foundations formwork shall be measured to the sides of such concrete foundations regardless of whether or not any formwork is used, except where it is expressly stated on the Drawings that the concrete is to be cast against the soil face.

**Itemisation**

3 Separate items shall be provided for formwork in accordance with Part II paragraphs 3 and 4 and the following:

| Group | Feature |
|---|---|
| I | 1 Formwork. |
|   | 2 Void formers. |
| II | 1 More than 300 mm wide horizontal or at any inclination up to and including 5° to the horizontal. |
|    | 2 More than 300 mm wide at any inclination more than 5° up to an including 85° to the horizontal. |

# Formwork

                3  More than 300 mm wide at any inclination more than 85° up to and including 90° to the horizontal.
                4  300 mm wide or less at any inclination.
                5  Curved of both girth and width more than 300 mm at any inclination.
                6  Curved of girth or width of 300 mm or less at any inclination.
                7  Domed.
                8  Void formers of different cross section.

III            1  Different classes of surface finish.
               2  Permanent formwork of different types.
               3  Void formers of different materials.

**Formwork**

**Item coverage**

4  *The items for formwork shall in accordance with The Preambles to Bill of Quantities General Directions include for:*
(a) *falsework, centring, fabricating, assembling, cutting, fitting and fixing in position and taking measures to produce the required shapes of concrete;*
(b) *forming cambers and falls;*
(c) *linings and taking measures to produce the required finish to the surfaces of the concrete;*
(d) *cutting and fitting around projecting members, pipes, reinforcement and the like;*
(e) *individual fillets, chamfers, splays, drips, rebates, recesses, grooves and the like of 100 mm total girth or less when measured overall the faces in contact with the concrete;*
(f) *maintaining in place until it is struck and allowing for any variation from the minimum period for striking arising from prevailing weather conditions;*
(g) *striking, taking down and removing;*
(h) *concrete provided in lieu of formwork to fill overbreak and working space.*

**Void Formers**

**Item coverage**

5  *The items for void formers shall in accordance with the Preambles to Bill of Quantities General Directions include for:*
(a) *fixing void formers against displacement during concreting operations;*
(b) *capping or blocking off ends;*
(c) *sealing ends and joints.*

**Patterned Profile Formwork**

**Units**

6  The units of measurement shall be:
    (i)  patterned profile formwork..................square metre.

**Definition**

7  The term 'patterned profile formwork' shall mean formwork designed to produce a concrete face with a specified patterned

Formwork 86

profile comprising ribs, corrugations, troughs or other patterns in relief.
Formwork with a specified regular patterns of formwork joints shall not be classified as patterned profile formwork.

**Measurement**

**8** The measurement shall be the flat undeveloped area of the patterned concrete shown on the Drawings or ordered by the Engineer but measured over the face of openings of 1 square metre or less and features described in (c) below.
Patterned profile formwork shall not be measured:
(a) to construction joints (whether shown or not on the Drawings) skewbacks, stunt ends, steppings, bonding chases and the like;
(b) to holes, ducts, pockets, sockets, mortices and the like, not exceeding 0·15 cubic metres each in volume;
(c) to individual fillets, chamfers, splays, drips, rebates, recesses, grooves and the like, not forming part of the pattern and of 100 mm total girth or less when measured overall the faces in contact with the concrete;
(d) to edge of blinding concrete 75 mm or less in thickness;
(e) to upper surfaces of concrete inclined at an angle of less than 15° to the horizontal.

**Itemisation**

**9** Separate items shall be provided for patterned profile formwork in accordance with **Part II paragraphs 3 and 4** and the following:

| Group | Feature |
|---|---|
| I | 1 Patterned profile formwork. |
| II | 1 Horizontal or at any inclination up to and including 5° to the horizontal.<br>2 At any inclination more than 5° up to and including 85° to the horizontal.<br>3 At any inclination more than 85° up to and including 90° to the horizontal.<br>4 Curved at any inclination. |
| III | 1 Different types. |

**Patterned Profile Formwork**

**Item coverage**

**10** *The items for patterned profile formwork shall in accordance with the Preambles to Bill of Quantities General Directions include for:*
(a) falsework, centring, fabricating, assembling, cutting, fitting and fixing in position and taking measures to produce the required shapes and patterns of concrete;
(b) forming cambers and falls;
(c) linings and taking measures to produce the required finish to the surfaces of the concrete;

## Formwork

(d) cutting and fitting around projecting members, pipes, reinforcement and the like;
(e) individual fillets, chamfers, splays, drips, rebates, recesses, grooves and the like of 100 mm total girth or less when measured overall the faces in contact with the concrete;
(f) maintaining in place until it is struck and allowing for any variation from the minimum period for striking arising from prevailing weather conditions;
(g) striking, taking down and removing.

# Section 15: Steel Reinforcement for Structures

**Units**

1 The units of measurement shall be:
   (i) bar and helical reinforcement................tonne.
   (ii) fabric reinforcement..................square metre.
   (iii) dowels..................number.

**Measurement**

2 The mass of plain bar reinforcement shall be calculated on the basis that the nominal density of steel is 0·00785 kilogrammes per square millimetre of cross sectional area per linear metre; the mass of deformed bar reinforcement shall be calculated as the nominal rolling mass of the reinforcement. Steel bar supports to reinforcement where shown on the Drawings shall be measured as reinforcement.

3 Fabric reinforcement shall be measured as the area of work covered, the BS reference being stated.

**Itemisation**

4 Separate items ahall be provided for steel reinforcement for structures in accordance with Part II paragraph 3 and 4 and the following:

| Group | Feature |
|---|---|
| I | 1 Bar reinforcement of nominal size 16 millimetres and under.<br>2 Bar reinforcement of nominal size 20 millimetres and over.<br>3 Fabric reinforcement of different BS references.<br>4 Dowels of different diameters and lengths.<br>5 Helical reinforcement. |
| II | 1 Mild steel.<br>2 High yield steel.<br>3 Stainless steel. |
| III | 1 Bars of 12 metres length or less.<br>2 Bars exceeding 12 metres in length but not exceeding 13·5 metres and so on in steps of 1·5 metres. |
| IV | 1 Bars threaded through holes in members. |

# Reinforcement

**Reinforcement**

**Item coverage**

5 *The items for reinforcement shall in accordance with the Preambles to Bill of Quantities General Directions include for:*
(a) *cleaning, cutting and bending;*
(b) *binding with wire or other material;*
(c) *supports and spacers (except for steel bar supports to reinforcement where shown on the Drawings);*
(d) *extra fabric reinforcement at laps;*
(e) *welding.*

**Dowels**

**Item coverage**

6 *The items for dowels shall in accordance with the Preambles to Bill of Quantities General Directions include for:*
(a) *drilling holes or forming pockets, casting and grouting;*
(b) *protective caps, sleeves and wrappings.*

# Section 16: Concrete for Structures

**In situ Concrete**

Units
1 The units of measurement shall be:
(i) in situ concrete.................cubic metre.

Measurement
2 No deduction shall be made for:
(a) holes, ducts, pockets, sockets, mortices and the like not exceeding 0·15 cubic metres each in volume;
(b) reinforcement;
(c) individual fillets, chamfers, splays, drips, rebates, recesses, grooves and the like of 100 mm total girth or less when measured overall the faces in contact with the formwork;
(d) in the case of concrete with a patterned profile face, any indentations of 100 mm total girth or less when measured overall the faces in contact with the patterned profile formwork.

Itemisation
3 Separate items shall be provided for in situ concrete in accordance with Part II paragraphs 3 and 4 and the following:

| Group | Feature |
|---|---|
| I | 1 In situ concrete of different classes. |
| II | 1 Different types of cement. |
| III | 1 Blinding concrete 75 mm or less in thickness. |

In situ Concrete

Item coverage
4 *The items for in situ concrete shall in accordance with the Preambles to Bill of Quantities General Directions include for:*
(a) *trial mixes (for Specification Table 16/1 classes of concrete only);*
(b) *mixing, placing in or against any surface, including soil faces where required, compaction and surface finishing;*
(c) *curing and protection;*
(d) *formwork (as Section 14 paragraphs 4 and 10) to upper surfaces inclined at an angle of less than 15° to the horizontal;*
(e) *materials and attendance for sampling and testing carried out by the Engineer;*
(f) *falls, cambers, and shaped profiles;*
(g) *weep pipes, pipe sleeves and the like;*
(h) *construction joints (whether shown or not on the Drawings), skewbacks, stunt ends, steppings, bonding*

# Concrete

        *chases and the like including formwork (as Section 14 paragraphs 4 and 10), water bars and stops;*
- (i) *holes, ducts, pockets, sockets, mortices and the like not exceeding 0·15 cubic metres each in volume including formwork (as Section 14 paragraphs 4 and 10);*
- (j) *formwork (as Section 14 paragraphs 4 and 10) to edge of blinding concrete 75 mm or less in thickness;*
- (k) *concrete to fill over break and working space.*

## Precast Members

**Units**

5 The units of measurement shall be:
- (i) precast members, slabs, segmental units (excluding culverts), hinges (except the sealing of hinges), specially moulded blocks.................number.
- (ii) precast copings, plinths and the like of uniform cross section, culverts..................linear metre.

**Definitions**

6 The term 'precast' applies to any concrete unit cast on Site but not in its final position, and to concrete units manufactured off the Site.

**Measurement**

7 The measurement of culverts shall be the length measured at the invert level.

**Itemisation**

8 Separate items shall be provided for precast members in accordance with Part II paragraph 3 and 4 and the following:

| Group | Feature |
|---|---|
| I | 1 Precast members, slabs, segmental units, hinges and specially moulded blocks.<br>2 Precast copings and plinths.<br>3 Precast culverts. |
| II | 1 Different types. |
| III | 1 Different sizes. |
| IV | 1 Curved. |

**Precast Members, Slabs, Segmental Units, Hinges, Specially Moulded Blocks, Copings, Plinths and Culverts**
**Item coverage**

9 *The items for precast members, slabs, segmental units, hinges, specially moulded blocks, copings, plinths and culverts shall in accordance with the Preambles to Bill of Quantities General Directions include for:*

- (a) *trial mixes (for Specification Table 16/1 classes of concrete only);*
- (b) *concrete strength test results and supplying one copy to the Engineer;*

## Concrete

(c) formwork (as Section 14 paragraph 4);
(d) reinforcement (as Section 15 paragraph 5);
(e) surface finishing and curing;
(f) individual fillets, chamfers, splays, drips, rebates, recesses, grooves and the like;
(g) holes, ducts, pockets, sockets, mortices and the like including formwork;
(h) matching members as required for side by side construction;
(i) marking members for identification and delivery in matching sequence;
(j) lifting devices including removal and bearing plates;
(k) temporary bracing or stays to prevent displacement;
(l) bedding, jointing and pointing including cramps, dowels or other fixing devices;
(m) caulking and sealing between and under units and members including formwork;
(n) infilling to joints between adjacent units and members where the maximum width of the joint is less than 100 mm including formwork;
(o) cutting and trimming;
(p) in the case of the precast prestressed members and the like, tendons (as Section 17 paragraph 6) and stressing (including partially stressing) and grouting internal tendons (as Section 17 paragraph 7);
(q) in the case of precast and precast prestressed members and the like for incorporation in in situ post tensioned prestressed construction, forming, installing and sealing tendon ducts to profile; forming recesses in the concrete for anchorages and jack seatings; bearing plates; reinforcing helices, grout inlets, vents and other components including casting in;
(r) in the case of culverts, fittings, and dealing with existing flows.

### Treatment to Concrete Faces after the Striking of Formwork

**Units**

10 The units of measurement shall be:
    (i) treatment to concrete faces after the striking of formwork.................square metre.

**Measurement**

11 The measurement shall be the face area of the treated surface shown on the Drawings or ordered by the Engineer, except for concrete faces formed with patterned profile formwork in which case the measurement shall be on the same basis and be the same area as the patterned profile formwork required to produce that concrete face irrespective of whether the whole of that face is to be treated or not.

Concrete

**Itemisation**

**12** Separate items shall be provided for treatment to concrete faces after the striking of formwork in accordance with Part II paragraphs 3 and 4 and the following:

| Group | Feature |
|---|---|
| I | 1 To concrete faces after the striking of formwork. |
| II | 1 Bush hammering.<br>2 Exposed aggregate.<br>3 Knocked rib.<br>4 Other stated treatments. |

**Treatment to Concrete Faces after the Striking of Formwork**
**Item coverage**

**13** *The items for treatment to concrete faces after the striking of formwork shall in accordance with the Preambles to Bill of Quantities General Directions include for:*
(a) *removing loose material from the concrete and washing clean;*
(b) *particular requirements relating to the constituents of the concrete mix and its placing;*
(c) *in the case of exposed aggregate finish, the coating of the formwork with retarding agent and subsequent treatment of the surfaces upon removal of the coated formwork to obtain the required finish;*
(d) *construction and subsequent demolition of sample panels of facework.*

# Section 17: Insitu Post-tensioned Prestressing for Structures

**Units**

**1** The units of measurement shall be:
  (i) tendons, stressing and grouting, protective covering to external tendons................number.

**Definition**

**2** For the purpose of this Section a tendon is defined as all the permanent components of a system which imparts a compressive load to a concrete member through a common anchorage or bearing plate.

**Measurement**

**3** Lengths of tendons shall be measured along the line of the tendon between the outside faces of those parts of the anchorage units cast into the concrete. Tendons shall be grouped so that no member of the group differs in length from the stated length by more than 5%.

**4** Protective covering to external tendons shall be measured irrespective of whether the protection is applied at Site or at the place of manufacture.

**Itemisation**

**5** Separate items shall be provided for in situ post tensioned prestressing for structures in accordance with Part II paragraphs 3 and 4 and the following:

| Group | Feature |
|---|---|
| I | 1 Tendons.<br>2 Stressing and grouting internal tendons.<br>3 Stressing external tendons.<br>4 Final stressing and grouting tendons of members supplied partially prestressed.<br>5 Protective covering to external tendons. |
| II | 1 Tendons for in situ concrete construction.<br>2 Tendons for segmental construction. |
| III | 1 Tendons of different types.<br>2 External protective covering of different types or size. |
| IV | 1 Longitudinal tendons.<br>2 Transverse tendons.<br>3 Tendons in any other direction. |
| V | 1 Tendons of a differing stated length. |

Prestressing 95

**Tendons**

**Item coverage**

6 The items for tendons shall in accordance with the Preambles to Bill of Quantities General Directions include for:
(a) forming, installing and sealing tendon ducts to profile or between precast segmental units;
(b) steel cables, wires or strands with couplers, binders, spacers and proving that tendons are free to move between anchorages in ducts;
(c) tendon anchorages, bearing plates, reinforcing helices, grout inlets, vents and other components except where these are supplied complete with precast members or segments;
(d) forming recesses in the concrete for anchorages and jack seatings;
(e) allowing for variations of length in tendons contained in the same bill item;
(f) cleaning ducts.

**Stressing and Grouting Internal Tendons, Stressing External Tendons and Final Stressing and Grouting Tendons of Members Supplied Partially Prestressed**

**Item coverage**

7 The items for stressing and grouting internal tendons, stressing external tendons and final stressing and grouting tendons of members supplied partially prestressed shall in accordance with the Preambles to Bill of Quantities General Directions include for:

(a) checking the accuracy of load measuring equipment and adjusting;
(b) applying pre-stress in one or more stages;
(c) gripping and trimming tendons;
(d) taking observations and compiling a record of stressing and grouting operations and supplying one copy to the Engineer;
(e) in the case of internal tendons, grouting, sealing vent holes and end anchorages, treating ends of tendons and filling anchorages and jack seating recesses with in situ concrete (as Section 16 paragraph 4);
(f) accommodating and adjusting for differences between tendons included in the same bill item;
(g) calculation in respect of the required jacking force and extension;
(h) releasing tension and re-tensioning where pull-in is greater than that agreed by Engineer.

**Protective Covering to External Tendons**

**Item coverage**

8 The items for protective covering to external tendons shall in accordance with the Preambles to Bill of Quantities General Directions include for:
(a) tying or bonding to main structure;
(b) sealing at joints.

# Section 18: Steelwork for Structures

**Units**

1 The units of measurement shall be:
  (i) fabrication, permanent erection.............tonne.
  (ii) trial erection at the place of fabrication............
  ............item.

**Measurement**

2 The measurement shall be the total weight of the finished member comprising plates, rolled sections, shear connectors, stiffeners, cleats, packs, splice plates and all fittings computed in accordance with BS 153 Part 2, without allowance for tolerance for rolling margin and other permissible deviations from standard weights, and excluding the weights of weld fillets, bolts, nuts, washers, rivet heads and protective coatings. No deductions shall be made for notches, holes and the like each less than 0·01 square metre measured in area.

3 Deck panels shall be measured separately only when the deck panel is not integral with the main member. Bracings, external diaphragms and the like shall be measured separately as subsidiary steelwork only when they are not integral with main members or deck panels.

4 Main members and deck panels shall be inclusive of connectors, stiffeners, internal diaphragms and other integral components.

**Fabrication of Steelwork**

**Itemisation**

5 Separate items ahall be provided for fabrication of steelwork in accordance with Part II paragraphs 3 and 4 and the following:

| Group | Feature |
|---|---|
| I | 1 Fabrication. |
| II | 1 Main members.<br>2 Deck panels.<br>3 Subsidiary steelwork. |
| III | 1 Rolled sections.<br>2 Plated rolled sections.<br>3 Plated girders.<br>4 Box girders. |

## Steelwork

| | | |
|---|---|---|
| IV | 1 | Comprised of different combinations of BS 4360 grades of steel. |
| V | 1 | Curved section. |
| VI | 1 | Tapering section. |

**Fabrication**

**Item coverage**

6 The items for fabrication shall in accordance with the Preambles to Bill of Quantities General Directions include for:
(a) preparation and supply of detailed working drawings;
(b) examining and checking steel plate for segregation, laminations, cracks and surface flaws and carrying out any remedial measures required by the Engineer in respect of such defects;
(c) cutting, marking off, drilling notching, machining, form fitting, edge preparation and cambering;
(d) welding, riveting, bolting as the case may be, assembling and pre-heating;
(e) rivets, bolts, nuts and washers required to fabricate the steelwork and to complete the erection and installation of steelwork on Site, together with spares and service bolts, drifts, draw up cleats and the like specified in BS 153 Part 2;
(f) welding shear connectors to steel members either at the place of fabrication or on Site and pre-heating;
(g) qualification tests of welders;
(h) production tests of welding during fabrication including non destructive testing;
(i) marking members for identification and delivery in matching sequence.

### Erection of Steelwork

**Itemisation**

7 Separate items shall be provided for erection of steelwork in accordance with Part II paragraphs 3 and 4 and the following:

| Group | Feature |
|---|---|
| I | 1 Trial erection at the place of fabrication.<br>2 Permanent erection. |
| II | 1 Different forms of steel sub-structure construction.<br>2 Different forms of steel superstructure construction. |

**Trial Erection at the Place of Fabrication**

8 The items for trial erection at the place of fabrication shall in accordance with the Preambles to Bill of Quantities General Directions include for:

Steelwork 98

| | |
|---|---|
| Item coverage | (a) temporary bracing or stays to prevent displacement including the provision and removal of temporary attachments;<br>(b) proving dimensions, cambers and profiles;<br>(c) match-marking members as required for permanent erection;<br>(d) dismantling;<br>(e) modifications and refitting of members as a result of the trial erection. |
| Permanent Erection | 9 The items for permanent erection shall in accordance with the Preambles to Bill of Quantities General Directions include for: |
| Item coverage | (a) temporary bracing or stays to prevent displacement including the provision and removal of temporary attachments;<br>(b) qualification tests of welders;<br>(c) permanent bolted and welded connections required on Site including the provision of preheat and shelters for welding;<br>(d) production tests of Site welding including non-destructive testing. |

### Corrugated Steel Structures

| | |
|---|---|
| Units | 10 The units of measurement shall be:<br>(i) corrugated steel structures (stating the length) . . . . . . . . . . . . . . . . number. |
| Measurement | 11 The length stated shall be the extreme length of the corrugated steel structure. |
| Itemisation | 12 Separate items shall be provided for corrugated steel structures in accordance with Part II paragraphs 3 and 4 and the following: |

| Group | Feature |
|---|---|
| I | 1 Corrugated steel structures. |
| II | 1 Different types. |
| III | 1 Different sizes. |
| IV | 1 Different thicknesses. |

| | |
|---|---|
| Corrugated Steel Structures | 13 The items for corrugated steel structures shall in accordance with the Preambles to Bill of Quantities General Directions include for: |
| Item coverage | (a) preparation and supply of detailed working drawings;<br>(b) cutting, marking off, drilling, notching, bevels, skews, |

## Steelwork

bends, edge preparation, cambering, riveting, bolting and fabricating;
(c) laps, rivets, bolts, nuts, washers and the like required to assemble the structure and complete the erection and installation of the corrugated structure on Site, together with spares and service bolts, drifts, draw up cleats and the like;
(d) protective treatment;
(e) marking members for identification and delivery in matching sequence;
(f) channels including casting in to line and level.

# Section 19: Protection of Steelwork against Corrosion

**Units**

1 The units of measurement shall be:
   (i) protective system.................square metre.
   (ii) desiccants, vapour phase inhibitors...............
        kilogramme.

2 The measurement shall be the surface area to be treated and the weight of desiccant or inhibitor required.

**Itemisation**

3 Separate items shall be provided for protection of steelwork against corrosion in accordance with Part II paragraphs 3 and 4 and the following:

| Group | Feature |
|---|---|
| I | 1 Protective System.<br>2 Desiccants.<br>3 Vapour phase inhibitors. |
| II | 1 Different types. |

**Protective System**

**Item coverage**

4 *The items for protective system shall in accordance with the Preambles to Bill of Quantities General Directions include for:*
(a) *specimen panels of blast cleaning;*
(b) *paint samples and despatching to testing authority;*
(c) *paint application procedure trials;*
(d) *testing;*
(e) *masking and other measures to protect adjacent untreated steelwork;*
(f) *joint fillers and sealing of bolted joints;*
(g) *preparing materials for application;*
(h) *preparation of surfaces and painting of steelwork at the place of fabrication and on Site;*
(i) *complying with any special requirements in respect of ambient conditions for the application of protective treatment and for intervals between successive operations and applications;*
(j) *stripe coats;*
(k) *obtaining the correct dry film thickness of paint or other coating.*

## Protection of Steelwork

**Desiccants and Vapour Phase Inhibitors**

**Item coverage**

5 *The items for desiccants and vapour phase inhibitors shall in accordance with the Preambles to Bill of Quantities General Directions include for:*
 (a) *placing the desiccants or inhibitors in containers in the quantities and positions required;*
 (b) *taking precautions to prevent its exposure to atmosphere other than in its working environment.*

# Section 20: Waterproofing for Structures

**Units**

1 The units of measurement shall be:
  (i) waterproofing systems.................square metre.

**Measurement**

2 The measurement shall be the area of structural surface covered by the system. No deduction shall be made for openings of 1 square metre or less.

**Itemisation**

3 Separate items shall be provided for waterproofing for structures in accordance with Part II paragraphs 3 and 4 and the following:

| Group | Feature |
|---|---|
| I | 1 Different waterproofing system. |
| II | 1 More than 300 mm wide horizontal or at any inclination up to and including 45° to the horizontal.<br>2 More than 300 mm wide at any inclination more than 45° up to and including 90° to the horizontal.<br>3 300 mm wide or less at any inclination.<br>4 Domed.<br>5 Applied by brush or spray to surfaces of any width and at any inclination. |

**Waterproofing Systems**

4 The items for waterproofing systems shall in accordance with the Preambles to Bill of Quantities General Directions include for:

**Item coverage**

(a) preparing, cleaning and drying surfaces to be waterproofed including levelling courses;
(b) priming, adhesive coats, undercoats and intermediate layers;
(c) laying to cambers, falls and crowns;
(d) protective layer where expressly required in the Contract or as a result of the Contractor's choice of waterproofing system;
(e) formwork (as Section 14 paragraph 4);
(f) additional basecourse or wearing course required as a result of the Contractors choice of waterproofing system;
(g) forming or cutting nibs, angle fillets, external angles, mitres, stops and the like;

## Waterproofing

(h) sealing the membrane at its edges and around interruptions and projections;
(i) cutting out and rectifying imperfections;
(j) forming joints and overlaps;
(k) making good up to abutting surfaces including cleaning and priming.

# Section 21: Bridge Bearings

**Units**  1 The units of measurement shall be:
(i) supply of bearings, installation of bearings..........
..........number.

**Itemisation**  2 Separate items shall be provided for bridge bearings in accordance with Part II paragraphs 3 and 4 and the following:

| Group | Feature |
|---|---|
| I | 1 Supply of bearings. <br> 2 Installation of bearings. |
| II | 1 Different types. |
| III | 1 Different sizes. |

**Supply of Bearings**  3 *The items for supply of bearings shall in accordance with the Preambles to Bill of Quantities General Directions include for:*

**Item coverage**  *(a) nuts, bolts, washers, dowels, protective caps, dust covers, sockets, sleeves, wrapping, adhesives and lubricants;*
*(b) marking bearings for identification purposes.*

**Installation of Bearings**  4 *The items for installation of bearings shall in accordance with the Preambles to Bill of Quantities General Directions include for:*

**Item coverage**  *(a) drilling holes or forming pockets in the structure and casting in of bolts, dowels and sockets;*
*(b) forming plinths including formwork (as Section 14 paragraph 4);*
*(c) setting and releasing locking mechanism;*
*(d) adhesives and epoxy mortar, cement mortar or grout;*
*(e) painting and greasing.*

# Section 22: Metal Parapets

**Units**

1 The units of measurement shall be:
(i) fabrication of metal parapets, installation of metal parapets.................linear metre.

**Measurement**

2 The measurement shall be the length of complete parapet.

**Itemisation**

3 Separate items shall be provided for metal parapets in accordance with Part II paragraphs 3 and 4 and the following:

| Group | Feature |
|---|---|
| I | 1 Fabrication of metal parapets.<br>2 Installation of metal parapets. |
| II | 1 Different types. |
| III | 1 Different heights. |

**Fabrication of Metal Parapets**

**Item coverage**

4 *The items for fabrication of metal parapets shall in accordance with the Preambles to Bill of Quantities General Directions include for:*
(a) *preparation and supply of detailed working drawings;*
(b) *joint fittings, expansion joints, base plates, sockets, anchorage assemblies, holding down bolts, nuts, washers, fastenings and adhesives;*
(c) *qualification tests of welders;*
(d) *production tests of welding during fabrication including non destructive testing;*
(e) *marking parapets for identification;*
(f) *protective system applied at the place of manufacture;*
(g) *connection pieces for attachment of safety fences.*

**Installation of Metal Parapets**

**Item coverage**

5 *The items for installation of metal parapets shall in accordance with the Preambles to Bill of Quantities General Directions include for:*
(a) *drilling holes or forming pockets in the structure and casting in bolts, sockets, base plates and anchorage assemblies;*
(b) *adhesives and epoxy or polyester mortar, cement mortar or grout;*
(c) *protective system applied on Site;*
(d) *making good any damage or defects in the protective system applied at the place of manufacture.*

# Section 23: Movement Joints for Structures

### Movement Joints to Bridge and Viaduct Decks

**Units**

1 The units of measurement shall be:
  (i) movement joints.................number.

**Definition**

2 The term movement joint to bridge and viaduct decks covers all types of permanent joints or hinge throats which allow expansion, contraction, shrinkage or angular rotations to take place.

**Itemisation**

3 Separate items shall be provided for movement joints to bridge and viaduct decks in accordance with Part II paragraphs 3 and 4 and the following:

| Group | Feature |
|---|---|
| I | 1 Movement joints. |
| II | 1 Different types or materials. |
| III | 1 Different lengths or perimeters. |
| IV | 1 Different gap widths. |

**Movement Joints**

**Item coverage**

4 The items for movement joints shall in accordance with the Preambles to Bill of Quantities General Directions include for:
(a) preparing and cleaning the various surfaces;
(b) installing or constructing the joint complete with any special fittings at kerbs, footpaths, ducts and the like in the required positions including the use of templates, guides and temporary devices to retain the joint in the correct position during construction;
(c) setting the joint as directed by the Engineer having regard to the mean temperature of the bridge or viaduct at the time of setting the expansion gap, and to other factors affecting movement;
(d) inserting and protecting joint filler material and sealing the joint including priming the surfaces to be sealed or applying adhesives or other special materials and inserting sealing material or preformed sealing strips;
(e) forming or cutting sealing grooves in road surfacings above joints or in facework and sealing;
(f) protective or other treatment to steel components including surface preparation and greasing;

## Movement Joints

(g) *installing and maintaining temporary ramps or other safeguards to protect the joint against damage during construction, and removing when no longer required.*

### Movement Joints other than to Bridge and Viaduct Decks

**Unit**

5 The units of measurement shall be:
  (i) joint filler..................square metre.
  (ii) joint sealant, water bar or water stop................ linear metre.

**Measurement**

6 The measurement of joint filler shall be the area of the surface covered. The measurement of joint sealant shall be the length of the joint on the external face of the sealant. The measurement of water bar or water stop shall be the length along the axis.

**Itemisation**

7 Separate items shall be provided for movement joints other than to bridge and viaduct decks in accordance with Part II paragraphs 3 and 4 and the following:

| Group | Feature |
|---|---|
| I | 1 Joint filler. <br> 2 Joint sealant. <br> 3 Water bar or water stop. |
| II | 1 Different types or materials. |
| III | 1 Different sizes. |
| IV | 1 Different thicknesses. |

**Joint Filler**

**Item coverage**

8 *The items for joint filler shall in accordance with the Preambles to Bill of Quantities General Directions include for:*
(a) *cutting and shaping;*
(b) *preparing and cleaning the various surfaces;*
(c) *applying adhesives or other special materials;*
(d) *applying or inserting filler.*

**Joint Sealant**

**Item coverage**

9 *The items for joint sealant shall in accordance with the Preambles to Bill of Quantities General Directions include for:*
(a) *preparing and cleaning the various surfaces;*
(b) *priming the surface of the joint;*
(c) *inserting the compound and sealing the joint;*
(d) *protecting joint filler.*

**Water Bar or Water Stop**

10 *The items for water bar or water stop shall in accordance with the Preamble to Bill of Quantities General Directions include for:*

## Movement Joints

**Item coverage**

(a) *cutting, notching, welding, fittings and jointing;*
(b) *cutting or terminating joint filler up to water bar or water stop;*
(c) *placing and casting in.*

# Section 24: Brickwork for Structures

**Units**

1 The units of measurement shall be:
   (i) brickwork.................square metre or item.

Note: Facings shall be measured as extra over brickwork except where brickwork is built entirely of facings. The use of 'Item' as the unit of measurement should be restricted to brickwork in relatively small quantities and alteration works.

**Measurement**

2 The measurement by the square metre shall be the superficial area of brickwork. No deduction shall be made for openings of 0·10 square metre or less.

**Itemisation**

3 Separate items shall be provided for brickwork for structures in accordance with Part II of paragraphs 3 and 4 and the following, except where item is used as the unit for brickwork when only one such item shall be provided:

| Group | Feature |
|---|---|
| I | 1 Brickwork in different types of bricks. |
| II | 1 Different thicknesses. |
| III | 1 Different bonds. |
| IV | 1 Curved on plan. |
| V | 1 With a battered face. |
| VI | 1 In walls.<br>2 In facework to concrete.<br>3 In arches.<br>4 In alteration work. |

**Brickwork**

**Item coverage**

4 The items for brickwork shall in accordance with the Preambles to Bill of Quantities General Directions include for:

(a) *jointing, pointing and fair-faced work, including rough and fair cutting;*
(b) *plinths, corbels, bull noses, chases, rebates, quoins, brick copings, string courses and the like;*
(c) *ties, dowels, cramps, joggles and the like including fixing;*

**Brickwork**

(d) centring and temporary supports;
(e) bonding into existing work;
(f) reinforcement (as Section 15 paragraph 5);
(g) filling the cavity between the brickwork and the backing;
(h) building in pipes, holdfasts, bolts and the like and forming small openings;
(i) construction and subsequent demolition of sample panels of brickwork;
(j) damp proof courses;
(k) removing loose material from the backing and washing clean.

# Section 25: Masonry for Structures

**Units**

1 The units of measurement shall be:
   (i) general masonry.................cubic metre or item.

Note: The use of item as the unit of measurement should be restricted to masonry in relatively small quantities and alteration works.
   (ii) copings, specially shaped and dressed string courses and the like.................linear metre.
   (iii) individual masonry blocks, features or stones.................number.

**Definition**

2 The term Masonry shall be taken for the purpose of this Section to include cast stonework and concrete blockwork.

**Measurement**

3 No deduction shall be made from the measurement for holes or voids of 0·15 cubic metre or less.

**Itemisation**

4 Separate items shall be provided for masonry for structures in accordance with Part II paragraphs 3 and 4 and the following, except where item is used as the unit for masonry when only one such item shall be provided:

| Group | Feature |
|---|---|
| I | 1 General masonry.<br>2 Masonry copings.<br>3 Specially shaped and dressed string courses of masonry.<br>4 Individual masonry blocks, features or stones. |
| II | 1 Different types. |
| III | 1 Different coursing. |
| IV | 1 Curved on plan. |
| V | 1 With a battered face. |
| VI | 1 In walls.<br>2 In facework to concrete.<br>3 In arches.<br>4 In alteration work. |

Masonry

VII    1 Copings, string courses and the like of different cross sectional area.
       2 Individual blocks, features or stones of different sizes and shape.

**Masonry**

**Item coverage**

**5** *The items for masonry shall in accordance with the Preambles to Bill of Quantities General Directions include for:*
(a) *dressed stone facings including in situ dressing;*
(b) *setting, bedding, jointing, coursing, raking, forming quoins, grouting, pointing and fair-faced work including rough and fair cutting;*
(c) *centring and temporary supports;*
(d) *bonding into existing work;*
(e) *dowels, cramps, joggles or other fixing devices including the sinkings and mortices therefor and running in;*
(f) *tying to concrete or other types of structure;*
(g) *surface treatment to masonry next to concrete;*
(h) *filling the cavity between the masonry and the backing;*
(i) *building in pipes, holdfasts, bolts and the like and forming small openings;*
(j) *construction and subsequent demolition of sample panels of masonry;*
(k) *damp proof course;*
(l) *removing loose material from the backing and washing clean;*
(m) *obtaining manufacturers certificate and supplying copy to the Engineer;*
(n) *formwork (as Section 14 paragraph 4);*
(o) *reinforcement (as Section 15 paragraph 5);*
(p) *in situ concrete (as Section 16 paragraph 4);*
(q) *forming grooves, rebates, recesses, stoolings and weatherings;*
(r) *marking for identification and delivery in any matching sequence.*

Section 26 is not taken up

# Section 27: Testing

### Pile Testing

**Units**

1 The units of measurement shall be:
   (i) establishment of pile testing equipment............
           ............item.
   (ii) load testing of piles..................number.

**Measurement**

2 Measurement of the establishment of pile testing equipment shall be for each visit to the site of a bridge, viaduct or other structure ordered by the Engineer. Measurement of load tests shall be the number of tests ordered by the Engineer.

**Itemisation**

3 Separate items shall be provided for pile testing in accordance with Part II paragraphs 3 and 4 and the following:

| Group | Feature |
|---|---|
| I | 1 Establishment of pile testing equipment.<br>2 Load testing of piles. |
| II | 1 Piles of different types. |
| III | 1 Trial piling as a separate operation in advance of the main piling.<br>2 Main piling. |
| IV | 1 Load tests of different specified load and/or different angles of application of load. |

**Establishment of Pile Testing Equipment**

**Item coverage**

4 *The items for establishment of pile testing equipment shall in accordance with the Preambles to Bill of Quantities General Directions include for:*
(a) *bringing test rigs, kentledge and associated equipment to the site of the bridge, viaduct or other structure and subsequently removing it.*

**Pile Tests**

**Item coverage**

5 *The items for pile tests shall in accordance with the Preambles to Bill of Quantities General Directions include for:*
(a) *setting up test rigs, kentledge, cable anchorages, anchor piles and reference bench marks;*
(b) *setting up, operating and maintaining instruments and apparatus required to complete the test;*
(c) *constructing temporary pile caps and subsequently stripping off;*

# Testing

(*d*) *applying and re-applying the test load and releasing;*
(*e*) *taking readings, measurements and observations of pile performance under test and recording and supplying one copy of the record of each test to the Engineer.*

## Practical Tests of Concrete for Structures

**Units**

6 The units of measurement shall be:
   (i) practical tests of concrete................number.

**Measurement**

7 The measurement shall be the number of practical tests of concrete ordered and accepted by the Engineer as giving satisfactory results.

**Itemisation**

8 Separate items shall be provided for practical tests of concrete in accordance with Part II paragraphs 3 and 4 and the following:

| Group | Feature |
|---|---|
| I | 1 Tests with different moulds. |
| II | 1 Tests with different arrangements of reinforcement. |

**Practical Tests of Concrete**

**Item coverage**

9 *The items for practical tests of concrete shall in accordance with the Preambles to Bill of Quantities General Directions include for:*
(*a*) *erecting and dismantling trial moulds;*
(*b*) *formwork (as Section 14 paragraph 4 and 10);*
(*c*) *reinforcement (as Section 15 paragraph 5);*
(*d*) *placing compacting and curing concrete and providing facilities and assistance to the Engineer during inspection;*
(*e*) *disposing of concrete on completion of test;*
(*f*) *tests ordered by the Engineer and not accepted as satisfactory.*

## Testing Precast Concrete Members for Structures

**Units**

10 The units of measurement shall be:
   (i) testing precast concrete members................ number.

**Measurement**

11 The measurement of testing precast concrete members shall be the number of tests ordered and accepted by the Engineer as giving satisfactory results.

**Itemisation**

12 Separate items shall be provided for testing precast concrete members for structures in accordance with Part II paragraphs 3 and 4 and the following:

Testing

| Group | Feature |
|---|---|
| I | 1 Testing precast concrete members. |
| II | 1 Different types. |
| III | 1 Different sizes. |
| IV | 1 Non destructive tests of different maximum test load.<br>2 Destructive tests. |

**Testing Precast Members**

**Item coverage**

13 *The items for testing precast concrete members shall in accordance with the Preambles to Bill of Quantities General Directions include for:*
(a) *transporting members to and erecting at the place of test, and subsequently removing;*
(b) *applying and re-applying the loading in any way specified;*
(c) *the cost of testing at the Contractor's option all other prestressed members manufactured simultaneously on the same casting line as that of a member found on test to be unsatisfactory;*
(d) *taking readings, measurements and observations of members under test and recording and supplying one copy of the record of each test to the Engineer;*
(e) *tests ordered by the Engineer and not accepted as satisfactory;*
(f) *in the case of destructive tests—supply of the member.*

**Grouting Trials**

**Units**

14 The units of measurement shall be:
    (i) grouting trials................number.

**Measurement**

15 The measurement of grouting trials shall be the number of trials ordered by the Engineer.

**Itemisation**

16 Separate items shall be provided for grouting trials in accordance with Part II paragraphs 3 and 4 and the following:

| Group | Feature |
|---|---|
| I | 1 Grouting trials with different duct sizes. |
| II | 1 Different duct configurations. |

**Grouting Trials**

**Item coverage**

17 *The items for grouting trials shall in accordance with the Preambles to Bill of Quantities General Directions include for:*
(a) *providing test specimens and mock-ups of ducts including tendons for the trials;*

# Testing

(b) *carrying out tests to determine the physical properties of the grout including bleeding;*
(c) *grouting test specimens, or mock-ups;*
(d) *dismantling and examining grouted ducts by breaking open or other means.*

## Welding and Flame Cutting Procedure Trials

**Units**

18 The units of measurement shall be:
    (i) welding procedure trials, flame cutting procedure trials................item.

**Measurement**

19 The measurement of procedure trials shall be the number of trials ordered and accepted by the Engineer as giving satisfactory results.

**Itemisation**

20 Separate items shall be provided for welding and flame cutting procedure trials in accordance with Part II paragraphs 3 and 4 and the following:

| Group | Feature |
|---|---|
| I | 1 Welding procedure trials at place of fabrication.<br>2 Flame cutting procedure trials at place of fabrication. |

**Procedure Trials**

**Item coverage**

21 *The items for procedure trials shall in accordance with the Preambles to Bill of Quantities General Directions include for:*
(a) *simulating the most unfavourable conditions liable to occur in the actual fabrication, with all combinations of steel grades and plate thickness;*
(b) *transverse tensile tests, transverse and longitudinal bend tests, impact tests, and macro examination tests;*
(c) *fillet weld fracture tests and macro examination tests;*
(d) *tests to determine the Vickers hardness value;*
(e) *trials ordered by the Engineer and not accepted as satisfactory.*

## Tests of Bearings

**Units**

22 The units of measurement shall be:
    (i) bearing tests..................number.

**Measurement**

23 The measurement of bearing tests shall be the number of tests ordered and accepted by the Engineer as giving satisfactory results.

**Itemisation**

24 Separate items shall be provided for test of bearings in accordance with Part II paragraphs 3 and 4 and the following:

Testing 118

| Group | Feature |
|---|---|
| I | 1 Compression tests.<br>2 Shear tests.<br>3 Bond tests.<br>4 Tests for physical properties.<br>5 Weathering tests. |
| II | 1 Bearings of different types. |
| III | 1 Bearings of different sizes. |

**Bearing Tests**

**Item coverage**

25 *The items for bearing tests shall in accordance with the Preambles to Bill of Quantities General Directions include for:*
(a) *taking readings, measurement and observations of the performance of bearings under test and recording and supplying one copy of the record of each test to the Engineer;*
(b) *tests ordered by the Engineer and not accepted as satisfactory.*

Section 28 is not taken up

# Section 29: Accommodation Works and Works for Statutory Undertakers

The method of measurement for Accommodation Works and Works for Statutory Undertakers shall be in accordance with the various Sections of this Method of Measurement or otherwise as a measured lump sum.